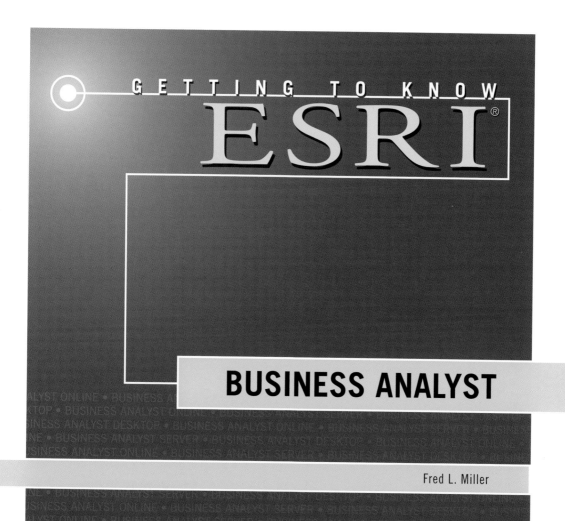

GETTING TO KNOW

ESRI®

BUSINESS ANALYST

Fred L. Miller

ESRI PRESS
REDLANDS, CALIFORNIA

ESRI Press, 380 New York Street, Redlands, California 92373-8100
Copyright © 2011 ESRI
All rights reserved.

15 14 13 12 11   1 2 3 4 5 6 7 8 9 10

Printed in the United States of America

Library of Congress Cataloging-in-Publication Data
Miller, Fred L., 1949-
    Getting to know ESRI business analyst / Fred L. Miller.
        p. cm.
    ISBN 978-1-58948-235-7 (pbk. : alk. paper) 1. Information storage and retrieval systems--Business.
2. Geographic information systems. I. Title.
    HF5548.2.M483 2010
    658'.0553--dc22   2010015540

Ask for ESRI Press titles at your local bookstore or order by calling 1-800-447-9778, or shop online at www.esri.com/esripress. Outside the United States, contact your local ESRI distributor or shop online at www.eurospanbookstore.com/ESRI.

ESRI Press titles are distributed to the trade by the following:

In North America:

Ingram Publisher Services
Toll-free telephone: (800) 648-3104
Toll-free fax: (800) 838-1149
E-mail: customerservice@ingrampublisherservices.com

In the United Kingdom, Europe, Middle East and Africa, Asia, and Australia:
Eurospan Group
3 Henrietta Street
London WC2E 8LU
United Kingdom
Telephone: 44(0) 1767 604972
Fax: 44(0) 1767 601640
E-mail: eurospan@turpin-distribution.com

CONTENTS

ACKNOWLEDGMENTS

Writing a book about Business Analyst is a bit like staying on the cusp of fashion. It must be done quickly and it doesn't stay done long. As you will learn in this book, the ESRI Business Analyst suite of products is an integrated set of six distinct software systems, each of which is constantly being updated and enhanced. In addition, annual data releases of current-year estimates and five-year projections of more than 1,500 demographic and lifestyle attributes, thousands of shopping centers, and millions of businesses complicate the process further.

Clearly, then, this book is not the product of one person, but rather a team of dedicated professionals committed to producing a timely, accurate, useful product. Members of that team include professionals at ESRI Press, the company's Industry Solutions Group, and the Business Analyst Products Group.

Judy Hawkins, ESRI Press acquisitions editor, and Simon Thompson, ESRI's Industry Solutions retail and commercial team leader, saw the need for a book covering the rapidly growing field of business GIS in general and the Business Analyst suite in particular. Peter Adams, ESRI Press manager, and Mark Berry, Business Analyst products operations manager, concurred in that judgment and have supported the project enthusiastically.

The point person for the project was Michael Schwartz, project editor, who kept us all on track with patience and persistence. His dedication to the project and ability to balance the roles of a diverse team of very busy professionals were critical to our success.

David Boyles, editorial supervisor, helped coordinate the team's effort and provided invaluable editorial advice. Kelley Heider, Press projects coordinator, contributed to several editorial decisions and has been heavily involved in the distribution and marketing of the book. The ESRI Press production team converted our manuscript into the finished book you have in your hands.

The Industry Solutions and Business Analyst Product groups were instrumental in supporting the technical components of the project. Kristen Carroll, Business Analyst support specialist, served as technical consultant, fine tuning our descriptions of Business Analyst procedures, pointing us to additional functionality, assembling our software installation/resources appendix, and sharpening the text with insightful comments. At various points in the process, Business Analyst product engineers Jeff Hincy and Kyle Watson, product manager Bob Hazelton, lead product manager James Killick, and business group technical specialist Brad McCallum, assisted with technical questions. Support specialists Gregory Ponto, Randall Williams, Avi Cueva, and Andrew Stauffer guided us through some thorny patches. As always, Brent Roderick of Product Marketing provided invaluable assistance with the Tapestry Segmentation system, a vital component of this book. Similarly, Lucy Guerra, product manager for Address Coder, provided valuable advice on that product and the Segmentation Module as well.

The efforts of these professionals enhance every page of this book. I am deeply grateful for their contributions to this book and through it, dear reader, to you.

**This book is dedicated to Luther Adem Holbrook Miller
and any brothers, sisters, or cousins who lie in his future.**

Preface

The premise of *Getting to Know ESRI Business Analyst* is simple. Integrated business GIS in general and the ESRI Business Analyst in particular are no longer solely the tools of specialists in large organizations, but have evolved into essential information technology resources for enterprises of all types and sizes. The objective of this book is to help current and future professionals in this broader arena, including small-to-medium-business owners, understand and exploit these technologies.

It does so by tracing the development of a hypothetical start-up business from its original concept to its emergence as a national retail and service enterprise. Each chapter describes a new stage in that growth process and guides readers in using the tools in the Business Analyst to support it. As readers develop their Business Analyst skills, they help the enterprise achieve its growth and expansion goals.

This structure is a useful learning approach for three reasons:

First, the Living in the Green Lane scenario (initially presented in the Introduction) provides a consistent context that illustrates the matching of Business Analyst resources with an organization's evolving situation and objectives. Although the scenario provides context, it is also possible for readers wishing to concentrate on the procedural dimensions of the software to use the exercises directly. Thus, readers seeking guidance on how to perform specific tasks in Business Analyst can refer directly to the relevant sections of the book without referencing the Living in the Green Lane scenario.

Second, although the development of Business Analyst applications within this scenario is logically sequential, the exercises are *not* operationally sequential. That is, each chapter contains all the resources necessary for completing the tasks it describes. Readers are *not* required, for example, to complete chapter 4 in order to have the appropriate map files, datasets, and analyses necessary for chapter 5. Thus, readers who wish to focus on the chapters of most relevance to their interest may do so easily.

Third, this pedagogical structure is effective for many groups of learners. Students can use it to develop increasingly valuable and widely recognized career skills. Business professionals can use it to assess the potential for integrated business GIS in their companies. Entrepreneurs can use it to learn how to use integrated business GIS technologies as they develop business models and marketing plans for their enterprises. Economic development professionals can use it to explore the value of integrated business GIS as a support tool for new and existing clients.

As a learning resource, this book may be used as:

An application resource for a dedicated business GIS course

Getting to Know ESRI Business Analyst provides the practical exercises for a dedicated business GIS course. It assumes readers have mastered basic GIS skills and have some understanding of marketing. This course might be an advanced course in a marketing, geography, entrepreneurship, or information technology program. Increasingly, it might also be an advanced course with a business GIS emphasis, a certificate, and minor or major degree program.

This tutorial book supports the software application component of the course. It is designed to be integrated with a textbook or other reading materials and—ideally—a practical Business GIS research project for a client organization that uses the resources of Business Analyst.

An application resource for training programs and/or professional development seminars

Getting to Know ESRI Business Analyst is intended to be the core resource in programs designed to develop integrated business GIS skills within the context of professional development. For example, enterprises planning to use ESRI Business Analyst Server across the organization might use this book to develop new users' skills with the technology. Economic development, entrepreneurship, and/or business incubator organizations might use the book in seminars to enhance clients' understanding of the technology and skill in its use.

In this context, the book provides practical exercises that could be matched to the anticipated applications of seminar participants. Entrepreneurship seminars, for example, might focus on early environmental scanning and site-selection chapters. Professional development seminars within mature companies might focus more on the customer profiling, segmentation, and market-expansion applications of later chapters. Organizations training new users of Business Analyst Server resources could combine that chapter with those covering the specific applications users will be performing.

A self-study resource for professionals who wish to understand the potential benefit of integrated business GIS for their organizations and/or their clients

This approach is appropriate for learners who wish to enhance their understanding of the value of integrated business GIS and/or to develop business GIS skills on their own. These readers can select the chapters most relevant to their professional interests and concentrate their attention there. For them, the remaining chapters might serve to explain potential options for extending the use of Business Analyst in their organizations in order to increase the return on their investment in this technology.

In fact, the book provides a menu of options for developing the most relevant integrated business GIS skills as well as planning the process for future deployment of additional Business Analyst resources.

Software and data configurations

The core of this book is a series of hands-on exercises with all the major functions of the ESRI Business Analyst suite of products. These exercises require access to ArcGIS 9.3.1 software, Business Analyst Desktop 9.3.1, and the Segmentation Module 9.3.1. In addition, chapter 1 requires a Business Analyst Online premium subscription.

Readers with these software resources already installed and an established Business Analyst Online premium account may use the book directly. These readers need only import the LITGL Minneapolis St Paul.zip project file from the enclosed DVD into their Business Analyst system using the Project Explorer import function. This compressed project file can also be downloaded from the book's Web site at `www.esri.com/esripress`.

To allow readers who do not already have an account to perform the exercises in chapter 1, the book comes with a 30-day trial premium subscription to Business Analyst Online, which is available at `www.esri.com/gtkba`.

ESRI also offers a variety of licensing options to meet the needs of students, faculty, and educational institutions of all sizes. These licenses are available for academic, research, or administrative use by individuals, labs, departments, or entire campuses. Commercial use is prohibited for all academic licenses. For further information call 800-447-9778 or e-mail `academicsales@esri.com` with the subject Academic Licenses Inquiry.

These alternatives will provide readers with a comprehensive overview of Business Analyst's capabilities and extensive hands on exercises with its major tools. With these resources, readers can experience the power of the Business Analyst suite and evaluate its potential contribution to decision making in their enterprises.

INTRODUCTION

Business geographic information systems (GIS) provide powerful tools for managers wishing to use spatial tools to understand their organizations and operations more completely. The emergence of integrated business GIS solutions (IBGIS) enables enterprises of all sizes to exploit spatial analysis in more effective decision making. Enterprises that exploit this resource gain competitive advantage through a better understanding of their competitive environments, markets, customers, operating processes, and growth opportunities.

This overriding reality above marks an evolution in the development of GIS technologies in the business world. Once the province of a relatively small group of specialists skilled in the minutiae of arcane software running on powerful, graphics-intensive computer systems, GIS has emerged as an enabling technology for analysts, researchers, managers, and problem solvers in a variety of disciplines. In the business world, this has meant broader dissemination of GIS tools, the development of specialized analytical models and procedures for business problems, and the design of interfaces for managerial users with relatively little formal GIS training. Indeed, the merging of these trends into powerful integrated business GIS solutions provides the focal point for this book.

In the broadest sense, geographic information systems are families of information technology tools that support spatial inquiry and reasoning. That is, they enable users to pose, research, and analyze spatial questions as well as to communicate spatially relevant information in decision-making processes. GIS systems include computer hardware and software components, data aggregation and integration capabilities, professionals with specialized skills in spatial analysis, and information clients (managers and executives in the business GIS framework) with problems to be solved and decisions to be made.

Once confined to the desktops of specialists, GIS resources are now ubiquitous across the information technology landscape. GIS capabilities have been extended to Web-based applications, integrated into commercial and enterprise systems, packaged as server applications for internal and external organizational clients, and offered as spatial building blocks—or objects—for developers to use in designing enterprise solutions.

All this is in the organizational information technology (IT) world. In the broader world, consumer-oriented mapping systems such as ArcGIS Explorer, MapQuest, Google Maps, Google Earth, and Microsoft Bing Maps have educated millions of users in the richness of spatial and spatially organized information. More importantly, these systems have familiarized users with the procedures for using map-based interfaces to organize and explore information. They then are able to navigate map-based interfaces more competently when they encounter them at work or on organizational Web sites.

GIS tools are used in a wide range of environments, organizations, and applications. When applied to business decision making, they are known as business GIS. The general benefits of GIS resources in business decision making are well established. GIS systems enable business professionals to view the geographic dimension of business data directly, to understand it more fully, to perceive relationships with other spatial information more effectively, to resolve spatial problems in business operations more comprehensively, and to communicate the rationale for those decisions more clearly. [1]

Despite this rich potential, businesses have been relatively slow to adopt GIS technologies. The constraints to more rapid adoption include:

• Hardware and software acquisition and maintenance costs
• Data acquisition and management costs
• Human resource costs to acquire and maintain GIS skills
• Limited awareness of GIS capabilities among information technology professionals [2]

The emergence of integrated business GIS solutions has allowed organizations of all sizes to exploit the capabilities of business GIS while overcoming many of these constraints. These solutions are built on standard GIS platforms with four very significant enhancements. The first enhancement is the bundling of extensive commercial data from a variety of sources into integrated systems. This streamlines data acquisition, lowers overall data costs, and addresses the issue of maintaining current data for business GIS analysis.

The second enhancement is the ability to interface with organizational IT systems to integrate enterprise data into business GIS systems, join it with bundled commercial data, and analyze it to reveal spatial patterns relevant to business decisions. This capability of creating new insights into organizational data through spatial analysis is a key benefit of business GIS.

Many of those insights come from the third enhancement of integrated business GIS solutions: a rich collection of automated tools, wizards, and procedures to perform advanced spatial analysis. By automating these procedures, organizing them logically, and presenting them to users in familiar, wizard-based interfaces, these solutions place powerful analytical tools at the disposal of users with relatively little GIS training. This allows greater access to these tools by managers and supports more extensive use of spatial insight in business decision making.

The fourth enhancement consists of a wide variety of standardized reports, maps, and graphs that communicate the results of GIS analyses. These documents capture relevant information efficiently and present it clearly to enable more effective communication of spatial analysis. In this way, they provide strong support for recommendations resulting from that analysis. In many IBGIS implementations, the formats for these documents are customizable, allowing users to produce exactly the type of reports and maps that best address the business issue under review.

Clearly, integrated business GIS solutions place significant GIS horsepower at the disposal of managers making business decisions. Though ripe with potential, these enhanced capabilities also pose challenges. To use them wisely, business managers must be familiar with the data bundled with the systems, the streamlined GIS tools available to them, the appropriate use of those tools, and the proper understanding of the reports, maps, and charts that they generate. Given the scope and power of these systems, this can be a daunting task. It becomes even more challenging when managers encounter a variety of integrated business GIS solutions options deployed throughout a range of their organizational information technology infrastructure.

This purpose of this book is to help managers meet this challenge. It focuses on one integrated business GIS solution: the suite of Business Analyst products offered by ESRI. To achieve this purpose, we will follow the life cycle of a single hypothetical enterprise from its inception to maturity. At each stage we will use Business Analyst tools to inform key business analysis and decision making. At each stage, we also will employ the most relevant business GIS tools, apply them to current commercial and organizational data, review the resulting documentation, and use this information in the decision-making process. To begin that process, let's learn a little more about this organization and its owners.

The Living in the Green Lane (LITGL) scenario

Stunned by the heating costs of Minneapolis winters, Janice Brown and her husband, Mark, decided to make the new home they were planning to build as energy efficient as possible. To meet that goal, Janice began an extensive research project on heating, cooling, and insulation options. She quickly learned that other systems in the home such as water, lighting, landscaping, energy, and water-saving appliances offered significant opportunities for increasing efficiency and lowering costs as well. Moreover, as the Browns' concern for environmental quality and the size of their collective "carbon footprint" grew, so did Janice's realization that their efforts at more efficient living likewise could decrease the environmental impact of their lifestyle.

As they built their new home, the Browns experienced difficulty in finding knowledgeable local contractors and builders who could help them implement the technologies they desired. Consequently, they developed relationships with several builders and contractors willing to take on new product lines and add new installation techniques to their portfolio of services. They did so in the belief that the demand for environmentally friendly homes and renovation projects would rise, and with it contracts for their newly developed services.

When the Browns' home was completed, it was featured in several area newspapers in articles that emphasized its distinctive design philosophy and economical operating costs. As a result, several families with similar interests contacted the Browns, as did contractors and builders wishing to emulate their approach. When she realized the growing interest in her research, Janice started to write a monthly newsletter that she sent to subscribers around the country. It highlighted new developments in green building practices and assessed the relative benefits of implementing each approach.

Demand for green building information soon grew beyond the scope of the newsletter format. Users were interested in more technical information as well as references to contractors and builders who could install the systems Janice wrote about. Many readers expressed interest in "green renovations," which improved the environmental quality and lowered operating costs of their existing homes, obviating the need to build new ones.

In response to this trend, Janice transformed her monthly newsletter into LivingInTheGreenLane .com,[3] an informational Web site offering a variety of services to users. These included reviews of emerging environmentally friendly technologies and products that included ratings of their effectiveness. Each week the site described a green building project, highlighting both new construction and renovation. Janice also used the site to answer readers' questions and provide advice on their own green projects. As the site developed, users asked increasingly detailed questions about specific brands of products and the skill, reliability, and expertise of the installers and contractors who worked with these products. Beginning with the contractors who worked on her home, Janice developed an extensive directory of professionals with experience in green products and projects. The directory included architects, designers, and builders, as well as general and specialized contractors.

As her circle of contacts in the green building movement expanded, Janice learned of the Minneapolis-St. Paul Green Builders Guild and its president, Steven Bent.[4] The Guild is a group of professionals in the construction industry whose primary goal is to promote green building practices. To qualify for membership, building professionals must be members of the U.S. Green Building Council,[5] a national organization of green builders, and have completed at least one Leadership in Energy and Environmental Design (LEED) certified house at the Silver level or above. The LEED for Homes certification system for home building assigns points for design and construction elements across eight categories for new homes, with certification levels at the Certified, Silver, Gold, and Platinum levels.[6] Guild members wish to promote green construction in the Minneapolis-St. Paul area but are constrained by the lack of a central location to demonstrate green building techniques and their benefits.

As Janice learned more about the Guild, she envisioned opportunities for expanding LivingInTheGreenLane.com in collaboration with the Guild and its members. She met with Steve to discuss the concept of expanding the LITGL concept to an actual retail store. The store would carry building materials, products, tools, and appliances consistent with LEED for Homes certification standards and procedures. In addition, it would include several booths and small workshop areas to support instructional activities. In these spaces Guild members would offer seminars, training, and information-sharing sessions to consumers interested in learning about green building and its potential. In one Living in the Green

Lane Home Center, customers could learn about green building approaches such as house design, renewable building materials, water-efficient plumbing systems, and energy-efficient heating and cooling systems, as well as proper siting and landscaping practices to limit external water use. Local Guild members in each of these fields would offer education, training, and maintenance services to help home customers integrate the full range of green resources into environmentally friendly, efficient home projects.

Janice and Steven believe that this set of goals requires a free-standing retail facility of 40,000 to 60,000 square feet. The facility must have substantial parking and space for outdoor demonstrations as well as shipping and inventory management capabilities. They prefer to acquire an existing building rather than construct a new one. This is consistent with LITGL's philosophy of improving the environmental performance of existing facilities whenever possible. It also provides the opportunity to demonstrate the value of green building practices by renovating a substantial existing facility to reduce its environmental footprint. Finding such a facility in the Minneapolis-St. Paul area that will meet these criteria, while providing a convenient shopping opportunity for green-conscious consumers, will be one of the critical "make or break" factors in their new enterprise.

Business GIS in a "born spatial enterprise"

To secure financing, Janice and Steven must convince banks and potential investors of the viability and profitability of their business model. To do so, they must write a business plan that provides a comprehensive description of this model, the customer base to which it will appeal, and the competitive environment in which it will operate. If their enterprise is successful, they must manage its growth, identify expansion opportunities, and extend its customer base. Each of these challenges has a significant spatial dimension. Understanding that dimension and using it to exploit opportunities more fully, serve customers more effectively, and communicate with investors more clearly will be key components of LITGL's success. Thus, Living in the Green Lane exemplifies the "born spatial enterprise," an organization which embraces business GIS as an integral part of its business processes from its inception.

The vehicle for this evolution is ESRI Business Analyst, a suite of products from the world's leading supplier of geographic information system (GIS) software. Business Analyst is an extension of the ArcGIS family of technologies. It integrates these technologies with extensive data collections, sophisticated wizard-based analytical tools, powerful reporting capability, and robust mapping tools. In short, Business Analyst is an integrated business GIS system that makes rich data and powerful analytic tools available to users with relatively modest GIS skills.

Business Analyst is available to users in several different formats. ESRI Business Analyst Online (BAO) is the Web-based version of the software which serves, as it will for LITGL, as an excellent point of entry into the business GIS world. ESRI Business Analyst Desktop (BA) extends its capabilities even further with extensive tools for trade-area analysis, site selection, customer profiling, site prospecting, sales territory design, and delivery/service routing applications.

The Segmentation Module add-on to Business Analyst Desktop allows users to understand customers more fully by integrating demographic information with lifestyle segmentation, purchasing, and behavioral data. This detailed portrait of existing customers can be used to improve service to them. More significantly, it can also be used to identify concentrations of households that match the profile of preferred customers. These concentrations present significant opportunities for market expansion.

ESRI Business Analyst Server expands the capability of organizations to integrate business GIS into their enterprise information technology systems. It allows IT and GIS professionals in these organizations to create business GIS workflows using Business Analyst technologies, integrate them into existing enterprise systems, and make them available to users throughout the organization. At this level ESRI Business Analyst is no longer a specialized tool used by a few experts in the organization. Rather, it is woven into the fabric of the enterprise, supporting the spatial awareness that is now an integral part of its analytical and decision-making processes.

This book will follow Living in the Green Lane throughout this process, from its initial business plan to its maturation as an organization with business GIS fully integrated into its enterprise information technology infrastructure. Each chapter will focus on a different stage in that process and on different products and technologies in Business Analyst. However, the central focus will be on the LITGL enterprise and how its business evolution determines its use of business GIS technologies.

The book is sequential in two ways. First, it follows the development of Living in the Green Lane from its inception to maturity as an enterprise. Second, it navigates through ESRI Business Analyst in a progression from the most broadly available Web-based systems (BAO) through desktop tools and extensions to enterprise-wide implementations of Business Analyst tools (BA Server).

That said, Living in the Green Lane illustrates only one of many possible paths to business GIS implementation. Other organizations may take different approaches. Many are not "born spatial enterprises" while others may wish to focus on individual applications rather than the holistic approach presented here. To support these readers, each chapter begins with an Executive Summary, which describes the business issue addressed in the chapter as well as the Business Analyst product and analytical tools employed. The Executive Summary also includes a listing of the cost and benefit factors relevant for the return on investment (ROI) assessment of the business GIS applications used in that chapter.

Taken together, the Executive Summaries can serve as a solution map for executives whose task is to decide which set of capabilities is most relevant for their organizations. In this approach, readers would identify and implement the most cost-effective method. When the implementation proves the ROI case to be solid, the organization then can move on to other applications in its evolution as a spatially aware enterprise.

To facilitate this flexibility in using applications in this book, exercises in each chapter are procedurally independent. That is, though the site-selection process in chapter 4 builds logically on the environmental analysis of chapter 3, readers can perform all the tasks in chapter 4 without completing those in chapter 3, and so on in subsequent chapters.

A final note: While this book covers a wide range of Business Analyst capabilities, it should be viewed as illustrative rather than exhaustive. Its purpose is not to serve as an operations manual, but to illustrate the role of business GIS in decision making throughout the business life cycle. As a reader, you should view your journey as an introduction to the capabilities of business GIS and Business Analyst rather than an inventory of those capabilities. As you work with these tools in the role of a Business GIS Analyst for Living in the Green Lane, be alert to other options within the software that might serve as more relevant tools for the issues facing your organization.

Preview of chapter content

With these considerations to guide you, here is the path that lies ahead:

Part I: Trade-area analysis and site reporting with Business Analyst Online

Janice and Steven's initial challenge is to secure funding for their new venture. To do so, they must write a business plan for investors that assesses the potential success of the enterprise within its consumer and competitive environments. This plan must reflect an understanding of these environments as well as include a business model for responding to them effectively. For retail enterprises such as Living in the Green Lane, store location is a critical success factor. Investors must be convinced that consumers in the Twin Cities will support the new store and be attracted to its location.

To make this case, Janice and Steven must describe the types of consumers who would be attracted to the store, identify concentrations of those prospective customers in the Twin Cities area, and evaluate alternative store locations from which to serve them. As their GIS planner, you will use thematic mapping, study area creation, and reporting capabilities of Business Analyst Online to perform these functions for LITGL. At the conclusion of this analysis, you will consider the return on investment (ROI) value of this application of BAO.

Chapter 1: Mapping the business environment: population and potential site characteristics

Janice and Steven begin their site-selection process by working with an economic revitalization agency in the Minneapolis-St. Paul area. That organization provides tax incentives for companies that acquire existing empty facilities in the area and use them for commercial purposes. To evaluate the benefits of this approach, you will use Business Analyst Online to define a study area and create thematic maps of its population characteristics. You then will define trade areas around this site using several different approaches, select the most appropriate trade-area model, and generate selected reports on its population characteristics. Finally, you will compare this area to other portions of the region to determine if more attractive sites are available. If that is the case, you will opt *not* to acquire the available revitalization site.

Part II: Business environment analysis with ESRI Business Analyst Desktop

As they continue the process of selecting a site for their first store, Janice and Steven must also consider the business environment in which they will operate and the opportunities for success within it. This is influenced not only by population characteristics of potential site locations, but also the competitive environment. That environment includes the proximity and size of competitors, the proximity and attractiveness of retail centers, and the transportation network that provides customer access to potential store locations. You will use the thematic mapping and map symbology capabilities of Business Analyst Desktop to identify and evaluate these factors. As these considerations are part of the initial site-selection process continued in Part III, the overall ROI analysis of this process is covered there.

Chapter 2: Thematic mapping with Business Analyst wizards and layer properties

Business Analyst Desktop provides automated tools for developing thematic maps and symbolizing various types of data layers. You will use these to map relevant demographic characteristics of the Minneapolis-St. Paul area. You will also use the settings available in Layer Properties to refine the maps you create, the symbology with which you represent data layers, and the classification schemes you use to portray population and business characteristics.

Chapter 3: Advanced thematic mapping and symbology, creating datasets, dynamic ring analysis

In addition to its standard thematic mapping tools, Business Analyst Desktop provides users with significant capabilities for creating new data layers, calculating new attributes, and customizing layer symbolization. These capabilities allow users to tailor their maps and data to the demands of specific research projects. Janice and Steven require that this type of data manipulation includes some characteristics of the green customer profile in their business environment analysis. You will perform these steps to create and map the data layers required to achieve this end. You will also use dynamic ring analysis to identify potentially attractive spots for the company's first store.

Part III: Trade-area analysis and site selection without customer data

Having integrated population data about the Minneapolis-St. Paul Core-Based Statistical Area (CBSA) with information about their competitive environment, Janice and Steven are ready to select a site for their first store. To do so, they will consider several trade-area models from Business Analyst Desktop, use one of them to define trade areas around six available sites, consult reports generated by Business Analyst Desktop to evaluate those sites, and select the most attractive one. At the conclusion of this process, you will consider the ROI impact of Business Analyst Desktop applications in retail site selection.

Chapter 4: Geocoding and evaluating alternative potential sites

A commercial real estate agent provides Janice and Steven with a list of six available retail locations in the Minneapolis-St. Paul area. They assign you to evaluate the attractiveness of these sites. You will begin by geocoding the addresses of the sites and displaying them on a map. You will then use Business Analyst Customer Prospecting and Trade Area Analysis tools to identify concentrations of attractive customers, create threshold rings of potential sales, and display trade areas of competitors. You will also use Locator Reports to quantify shopping centers and competitors in the vicinity of each available site. This information will inform the final site-selection decision you make in chapter 5.

Chapter 5: Defining trade areas, generating reports, selecting best site

Having analyzed the business environment, identified pockets of attractive customers, and assessed the competitive environment, you are ready to select the best site for the first Living in the Green Lane store. In this process, you will use Business Analyst Desktop to define specific trade areas for the available sites, generate reports on their characteristics, use this information to select the most appropriate site, and create map documents to support your recommendation.

Part IV: Customer profiling and site selection with customer data

Living in the Green Lane is successful in its basic green-building business, but Janice and Steven believe that customers would welcome a more comprehensive approach to green living, one that encompasses wellness values, preference for local organic foods, and participation in recycling and reuse programs. If this is true, Janice and Steven plan to broaden the LITGL concept from that of a "green-home center" to a "green-lifestyle center." They then plan to extend this concept to other parts of the Minneapolis-St. Paul metropolitan area by opening two additional stores to serve a population cluster that matches the profile of the most valuable customers from their first store.

Chapter 6: Building a profile of distinctive customer characteristics

Janice and Steven provide you with a list of Green Living Club members and direct you to create a profile of the best customers in this group and assess their affinity for the longer list of product/service lines they wish to implement in expanding the company's marketing concept. You will use the geocoding and spatial overlay functions of Business Analyst Desktop to attach demographic attributes to each customer record and to assign each customer to a Tapestry Segmentation segment. You then will use the Tapestry Segmentation data to learn more about customer preferences, purchasing patterns, media exposure, and lifestyle values with Market Potential Indexes. Finally, you will use this data to select the specific product/service line additions necessary to appeal to the green consciousness of customers with this profile.

Chapter 7: Customer-based trade-area analysis and site selection

The customer profile and expanded product/service mix you designed in chapter 6 provide the foundation for the site-selection process for Living in the Green Lane's second and third stores. You will use this customer data and the trade-area options of Business Analyst Desktop to define customer-derived trade areas around the existing store. You also will use the Find Similar function to identify areas of the Minneapolis-St. Paul area with concentrations of households that match this profile. In addition, you will use Principal Components Analysis to evaluate the potential of several alternative sites and an Advanced Huff Model to estimate distance decay and sales levels around these sites. These analyses will allow you to select two locations for new Living in the Green Lane stores in the Minneapolis St. Paul CBSA.

Part V: Sales territory management and route optimization

Living in the Green Lane's three stores initiate home services for pest control as well as lawn and garden maintenance services—all using organic products. A sales team of six representatives will sell the services and each store will have a service team to provide them. To manage these new services effectively, you will create a sales territory system that balances sales potential equitably among the three stores and six sales representatives. You will also use Business Analyst Desktop to optimize the routing of daily calls by one of the service teams.

Chapter 8: Sales territory design and balancing; route optimization

Living in the Green Lane's sales territory system requires two levels, with two sales representatives assigned to each of the three stores. You will use the Territory Design extension of Business Analyst Desktop to design an initial structure based on annual lawn and garden expenditures and total population. You then will refine this system by reassigning ZIP Codes between territories to correct geographic imbalance and align territories with existing transportation patterns.

In its initial implementation of its new service systems, LITGL will assign a service team to each store. Each team provides services to households in the store's sales territories. To maximize the efficiency of these operations and minimize drive time, you will use Business Analyst Desktop's routing tool to determine the optimal route for one day's service calls for a team, illustrating the potential savings of this technology when applied across the service routing function.

Part VI: Customer profiling and segmentation with the ESRI Business Analyst Desktop Segmentation Module

All three Living in the Green Lane stores are profitable and the company's green-lifestyle center concept has proven successful. Janice and Steven wish to expand the pace of growth significantly through increased sales to existing markets and expansion to new geographic markets. Their expansion strategy envisions a combination of company-owned stores and franchise agreements with local partners. As the company emphasizes relatively small, local service regions, they plan to enter each new market with at least four stores, of which at least two are company owned.

Both of these growth strategies require greater understanding of Living in the Green Lane's customer base. In the first case, a detailed profile of existing customers, their lifestyles, and their buying habits will help the company refine its marketing strategies to serve them more effectively and reach similar customers in current market areas. In the second case, the profile will serve as a model for evaluating marketing opportunities and projecting revenues in other geographic areas in the United States. You will use the Segmentation Module extension of Business Analyst Desktop to perform these analyses.

Chapter 9: Creating customer profiles

The Segmentation Module provides two approaches for creating customer profiles. The first is the Address Coder system, which is built into the module. Using it you will geocode the addresses of the 1,800-plus members of the Green Living loyalty club—LITGL's best customers.

With Address Coder you will also produce reports summarizing the demographic and lifestyle characteristics of these customers as well as a Business Analyst map layer.

The Segmentation Module also provides profiling capabilities for several population groups. You will create profiles of your customers and the Minneapolis-St. Paul CBSA. This will allow you to compare the characteristics of your best customers to the general population in which they live. What's more, the Segmentation Module allows you to create segments based on consumer survey data. This capability enables you to explore the characteristics of consumers who are, for example, heavy purchasers of organic lawn and garden products. The profiles you create in this chapter will provide the basic data for segmentation analysis in chapter 10.

Chapter 10: Segmentation analysis for enterprise expansion

In its segmentation analysis procedures, the Segmentation Module integrates customer profile data with internal enterprise sales data, Tapestry Segmentation lifestyle information, and Mediamark Research's consumer expenditure data to produce a comprehensive view of a company's customer base. You will use it to identify the most important groups of LITGL customers based on highly concentrated Tapestry Segmentation segments among Green Living members as well as those groups with the highest levels of average annual purchases.

Once these segments and their spending patterns are identified, you will develop LITGL's penetration strategy by exploring their purchasing, lifestyle, and media-exposure patterns and devising responsive marketing strategies. In addition, you will assess the level of LITGL's market penetration in its trade areas and perform gap analysis to identify geographic areas near existing stores with lower than expected customer households. These areas are attractive targets for the penetration strategy.

You will use the Segmentation Module to develop LITGL's expansion strategy as well. Using the Core, Developmental and Niche groups that you define, and the expenditure patterns of these segments, you will project the company's potential sales level in other CBSAs across the United States. You then will focus on an attractive CBSA and use trade-area assessment tools to determine if it will support the necessary four Living in the Green Lane stores required by Janice and Steven. Once appropriate CBSAs are identified, the trade-area and site-selection analyses you performed in Minneapolis-St. Paul would be repeated to ensure success in the new markets as well

Part VII: Expanding enterprise integrated business GIS with ESRI Business Analyst Server

With its national expansion, Living in the Green Lane has completed its growth from a local start-up company to a vibrant, growing national enterprise. Integrated business GIS has been a valuable tool at each stage of that growth process and plays a significant role in the day-to-day management of the company's operations. At this stage, it has grown beyond the scope of a single business GIS Analyst, even one with your advanced capabilities. It is necessary, therefore, to empower managers in each of the company's new markets and stores to utilize integrated business GIS tools in their operations and decision making.

The system that enables this capability is ESRI Business Analyst Server, a server-based business GIS system that allows analysts to aggregate relevant business GIS maps, data, and analyses, then deploy them across the Web for use by other managers or, in some cases, the company's customers. By combining central hosting of data, maps, and tools with substantial processing capacity in browser-based GIS clients, Business Analyst Server transforms the ESRI Business Analyst from a powerful analytical tool for individual analysts to more comprehensive system that supports enterprise decision making with business GIS capabilities at all levels of the enterprise.

Chapter 11: Serving applications with ESRI Business Analyst Server

In this chapter you will follow the process of disseminating integrated business GIS resources across the enterprise from both system administrator and user perspectives. For the administrator perspective, this process has three stages. In the first, enterprise GIS managers author and serve GIS resources to other users across the Web. This involves transforming the maps, analytical tools, and reporting functions from desktop activities to online mapping and processing services. Business Analyst Server builds on the underlying capabilities of ArcGIS Server to accomplish these tasks.

In the second stage, user roles and capabilities are defined for enterprise managers. While this task serves a security function, its primary goal is to match business GIS resources appropriately with the roles, responsibilities, and skills of the managers who will be using them. Thus, this function ensures that managers will be using the maps, data, procedures, and reports most appropriate for their decision-making responsibilities.

Once suitable roles and permissions are established to ensure proper access, the third stage of building role-specific workflows begins. Workflows are systematic streams of tasks structured to produce the analytics and metrics that support specific managerial decisions. Site selection and customer profiling are examples of the types of workflows implemented in Business Analyst Server. Business GIS managers may use the standard workflows provided in Business Analyst Server, customize those workflows, or design their own customized workflows to meet the specialized needs of the enterprise.

You also will work with Business Analyst Server from a user perspective. Specifically, you will assume the role of a new store manager charged with developing a profile of the store's customers to improve merchandising and to design integrated marketing communications to reach attractive new customers.

In this role, you will use the workflow developed in the first part of the chapter to geocode customers, map their location, identify their distinctive characteristics, identify Core and Developmental segments, and understand their purchase and media exposure patterns. These tasks will illustrate how Business Analyst Server tailors its business GIS resources to the needs of individual managers and how workflows can be used to standardize the analytical process and, over time, create a structured professional development path in which managers expand their business GIS skills as their professional responsibilities increase.

Part VIII: Conclusion

As a "born spatial enterprise," Living in the Green Lane has relied upon integrated business GIS for developing its marketing strategies, guiding its growth, and supporting its operations. As the company continues to evolve, so too will its reliance on Business Analyst applications.

Chapter 12: Growth trajectories with integrated business GIS

This chapter discusses the four ways in which this evolution will occur. The first is a more sophisticated use of existing Business Analyst tools. As LITGL's knowledge of its customers, competitors, and opportunities develops, its ability to exploit the advanced modeling and analytical functions of Business Analyst also will grow.

Second, integrated business GIS tools will become even more enmeshed in the company's daily operations. Web-based applications will support company-wide dissemination of effective customer and market analysis tools. Integration of GPS technologies deployed across mobile devices will increase the effectiveness of sales personnel and service technicians alike. Integration of Business Analyst with other enterprise information resources will support more effective exploitation of those resources and the valuable marketing data they contain.

Third, as Business Analyst develops, it will provide increasingly sophisticated tools for business GIS applications. Thus, more elaborate tools for, say, trade-area analysis, territory design, and customer segmentation will increase the value of these applications for Living in the Green Lane.

Finally, as the demands on modern enterprises continue to increase, so too will the ability of integrated business GIS to address them. Undoubtedly business environments are becoming increasingly risky and, in many ways, dangerous. Similarly, increasing concern for the physical environment and global climate change will result in expectations for greater accountability of business performance in these areas. Thus, applications such as Business Continuity Planning and Sustainability Assessment and Reporting will become more important. And, as these are inherently spatial analyses, they will necessarily entail expansion of integrated business GIS tools within enterprises that perform them.

In short, though this book offers a significant introduction to the value of integrated business GIS throughout the development of a business enterprise, its story is not complete. Indeed, it describes only the first steps of a continually evolving adventure with integrated business GIS as a core resource for making key business decisions. The most exciting part of that journey lies ahead.

Notes

1. Green, Richard P. and John C. Stager. 2005. Techniques and methods for GIS in business. In *Geographic Information Systems in Business*, James Pick, ed. Hershey, Pa.: Idea Group Publishing; Miller, Fred L. 2007. *GIS tutorial for marketing*. Redlands, Calif.: ESRI Press.

2. Keenan, Peter. 2005. Concepts and theories of GIS in business. In *Geographic information systems in business*, James Pick, ed. Hershey, Pa: Idea Group Publishing.

3. The people and Web site are fictional. Any resemblance to actual organizations and/or individuals is coincidental.

4. Both the person and the organization are fictional. Any resemblance to actual organizations and/or individuals is coincidental.

5. Von Paumgarten, Paul. 2003. The business case for high-performance green buildings: sustainability and its financial impact. *Journal of Facilities Management* 2 (1) June: 26–35.

6. Swope, Christopher. 2007. The green giant. *Architect* 96 (1) May: 134–37.

Part I
Trade area analysis and site reporting with ESRI Business Analyst Online

Relevance	Entrepreneurial new businesses need to develop responsive marketing plans to achieve success, and convincing business plans to secure funding.
Business scenario	Living in the Green Lane (LITGL) must write a business plan to secure financing for its first retail store. The plan must include a discussion of target customers, their attraction to the store, and projected sales.
Analysis required	LITGL must define its target customers, seek out concentrations of them in the Minneapolis-St. Paul area, estimate the size of the market near available sites, and clearly identify concentrations of favorable customer characteristics with thematic mapping.
Role of business GIS in analysis	Provides detailed reports on population characteristics, expenditures, and retail market patterns of variously defined market areas.
Integrated business GIS tool	ESRI Business Analyst Online.
ROI considerations: cost of business GIS	BAO subscription, training of researcher.
ROI considerations: benefits of business GIS	Increased sales from favorable location decision. Potentially lower loan costs if lender views the project as less risky based on customer data.

Table I.1 Executive summary

The Living in the Green Lane scenario

Janice Brown and Steven Bent must develop a business plan to persuade bankers and investors to support their Living in the Green Lane business initiative in the Minneapolis-St. Paul area. Their first task is to define a profile for green home-building consumers and seek concentrations of these households in the area. Then they must identify an appropriate site for a retail store to serve this target segment. Their preference is to renovate an existing retail facility to demonstrate the value of the green building techniques and products to be featured in their store.

Their desire to renovate an existing location is consistent with the objectives of the Twin Cities Redevelopment Task Force (TCRTF).[1] This nonprofit organization contributes to the area's economic development by renovating and reusing commercial facilities in the Twin Cities area. By offering tax incentives to organizations willing to invest in these properties, the Task Force seeks to stimulate economic activity while revitalizing the area's existing assets.

To support its objectives, the Task Force helps potential purchasers of these facilities with the market assessment and planning processes necessary to determine the financial feasibility of renovation. Business GIS is an important part of this service. The Task Force subscribes to Business Analyst Online and uses this resource to help potential purchasers assess market opportunities relative to the stock of available commercial properties. This is the resource Janice and Steven will use as they develop their business plan.

Janice's research on "green consumers" in the United States reveals segmentation patterns among consumers based on environmental interests. The 2007 GfK Roper Green Gauge Report identifies five distinct segments relative to environmental issues. The two segments with the highest level of environmental concern are True Blue Greens, with the strongest explicit commitment to environmental goals, and Greenback Greens, whose commitment is less ardent but who are willing to spend more on environmentally friendly products. Together these two segments comprised about 40 percent of U.S. households in 2007.[2]

Janice further discovered that, relative to home building, green consumers also divided themselves into three segments: Forest Greens, though relatively small in number, have a strong environmental commitment, and are willing to invest in green building practices solely for the sake of the environment. Greenback Greens are more cost conscious, but are willing to invest in green technologies that save them money. The most affluent of the three segments, Healthy Greens, are willing to invest in green building practices as part of an overall commitment to health and wellness. Demographically, Forest Greens and Healthy Greens tend to have higher levels of income, education, and home value than Greenback Greens.[3]

Based on this research, Janice has chosen income, education, and home value as the demographic factors she will use in defining Living in the Green Lane's target customers. With guidance from the Task Force staff, she will use Business Analyst Online to evaluate market opportunities based on these factors.

Business GIS tools in market assessment

To secure financing for their new enterprise, Janice and Steven must convince banks and potential investors of the viability and profitability of their business model. To do so, they must write a business plan that provides a comprehensive description of their business model, the customer base to which it will appeal, and the competitive environment in which it will operate. If their enterprise is successful, they must manage its growth, identify expansion opportunities, and extend its customer base. Each of these challenges has a significant spatial dimension. A key component of LITGL's success will be understanding that dimension and using it to exploit opportunities more fully, serve customers more effectively, and communicate with investors more clearly. Thus, Living in the Green Lane exemplifies what we will call a "born spatial enterprise," an organization that embraces business GIS as an integral part of its business processes from its inception.

The vehicle for this evolution is ESRI Business Analyst, a suite of products from the world's leading supplier of geographic information system (GIS) software. Business Analyst is an extension of the ArcGIS family of technologies. It integrates these technologies with extensive data collections, sophisticated wizard-based analytical tools, powerful reporting capability, and robust mapping tools. In short, Business Analyst is an integrated business GIS system that makes rich data and powerful analytic tools available to users with relatively modest GIS skills. In her initial work with the Task Force, Janice will be using the Business Analyst Online version of this system.

The most useful GIS tools for Janice's initial study are:

- Thematic mapping to identify concentrations of households with attractive characteristics.
- Market area definition to estimate the region that LITGL could serve from potential sites.
- Market area profiling to reveal the demographic and lifestyle characteristics of the designated market areas.
- Competitive analysis to identify potential competitors and retail attractions in the designated market areas.

Thematic mapping

The first of these tools, thematic mapping, is a basic GIS tool that allows users to display attributes of features graphically on a map. It is widely used to display surface characteristics, land-use data, and zoning information, among many other attributes. In business GIS applications, it is commonly used—as it will be here—to display demographic information at various levels of geography across a region of interest. This allows users to visually and quickly understand the distribution of the thematic layer and its relation with other data layers.

The demographic data displayed is among the most current available. ESRI's Data Products Division releases current-year estimates and five-year projections of a wide range of population demographic, employment, and expenditure measures at several levels of geography. This data is derived from the most recent census data, augmented with additional information from a variety of sources, and adjusted to the various levels of geography at which values are provided. To assist Janice with her analysis, you—as Living in the Green Lane's new GIS planner—will use Business Analyst Online to explore the distribution of household income and home value attributes at various geographic levels in the Twin Cities area. You then will use these maps to select one possible site for the first LITGL store.

Market area definition

With the second tool, market area definition, you will decide how to estimate the trade area of a retail location. (You will discover several methods for performing this task in Business Analyst. Business Analyst Online offers several, including simple rings, donuts, and drive-time polygons.) All three share the fundamental assumption that the ability of a retail location to draw customers has a significant spatial dimension. This means that for most retail locations the bulk of the customer base will be concentrated spatially around the store. However, that spatial concentration can be analyzed and understood a variety of ways.

For example, the simple ring approach uses direct geographic distance as the measure of attractiveness. As its name suggests, it draws a set of rings around a site at distances that estimate the area a store would serve from that site. While the ring approach is easy to produce and presents a visually balanced trade area, it ignores transportation infrastructure as well as natural features that might affect access.

The donut approach shares this method of defining a site's market area, though it differs in how it organizes data on the characteristics of households within the area. Specifically, the donut approach reports distinct data for each concentric ring in the trade area. For example, if rings are set at levels of 2 and 4 miles, the donut approach will report population characteristics for households within 2 miles of the site and distinct population information for households located beyond the 2-mile ring but within the 4-mile ring. Thus no households from the inner ring are included in the measures reported for the outer ring.

By contrast, the simple ring system reports population information cumulatively so that data reported for the 4-mile ring includes *all* households within four miles of the site, including those reported in the 2-mile ring area. This approach is appropriate if relevant demographic characteristics generally are consistent across the market area. However, if thematic mapping or other data suggest differences in characteristics among rings in the market area, the donut approach is more appropriate. Both methods are useful for quick market area estimates or in areas where natural features or road conditions pose few barriers to site access.

Drive-time polygons, the third market area definition option provided by Business Analyst Online, offer an alternative perspective on spatial proximity. This method assumes that most customers will be driving to the store and that their proximity is best measured by the time it takes to get there rather than distance from the store. As you would suspect, drive-time polygons extend further along expressways and highways than along residential streets. In addition, they reflect the way natural features are traversed by streets and highways, such as bridges that cross rivers, or slower roads that cover sloping terrain. Clearly drive times are approximations that must be adjusted for weather, time of day, and traffic conditions. Nevertheless, this approach remains more sensitive to customer access patterns than do either of the ring-based methods.

Market area profiling

The third GIS tool you will use, market area profiling, provides an incredibly rich collection of information about households in the market areas you have designated. In theory, this is a simple procedure: identify the households in the designated market area and calculate the statistics relevant to your inquiry. In Business Analyst Online, this procedure is equally simple: you request reports containing the information you desire for the market areas in which you are interested.

But this simplicity is deceptive. Market area profiling reports require detailed behind-the-scenes estimates and calculations. First, estimates must be made of the households that lie within the market area. This is difficult because the geographic units for which population demographics are reported are polygons. It is extremely unlikely that the borders of a market area polygon coincide exactly with those of the geographies (census tracts, block groups, ZIP Codes) which lie within it. When these areas overlap, it is necessary to estimate the portion of the households in a bisected block group that lies within the market area.

In Business Analyst these estimates are made using a sophisticated block group apportionment method that employs an attribute-specific process to allocate data attributes to trade areas based on the proportion of the group's block points that are within the trade area. These are point level units that contain basic population and household counts. As point features, block points are more clearly within a market area or outside it. Therefore, the proportion of block points in a particular block group lying within the market area is used as the method for weighting the values of that block group assigned to the market area.

Once you determine the units within the market area, you must calculate market area values for demographic and lifestyle measures from those of the constituent geographies. In this aggregation process, you also must weight these values by the population and household counts described above. This ensures that the market area figures more accurately represent the area as a whole.

These estimates and calculations lie behind the extremely broad selection of market area reports available in Business Analyst Online. As you seek out attractive sites for LITGL's first store, you will use several of these reports.

Competitive analysis

The fourth business GIS tool you will use in this chapter is competitive analysis. In Business Analyst Online, you can extract competitive information in the form of reports covering businesses, shopping centers, traffic counts, crime reports, and marketplace characteristics within designated trade areas. These reports summarize information on enterprises within the boundaries of these areas. As these data sources generally are point features, it is relatively easy to determine whether they lie within a market area polygon or not. The reports contain information about the count, size, sales levels, and employment of the establishments within designated areas. Maps available as reports provide spatial information for these establishments as well.

These, then, are the business GIS tools most relevant to Janice's initial study of the Twin Cities market environment. Using Business Analyst Online, you will apply these tools to the task of identifying market opportunities for the LITGL store.

Chapter 1

Mapping the business environment: population and potential site characteristics

As a Web-based application, ESRI Business Analyst Online is the most widely accessible product in ESRI Business Analyst. It hosts the most recent demographic data available from ESRI as well as a rich collection of data from other sources. Although it provides basic market area models and thematic mapping capability, its most valuable feature is a wide selection of reports for user-defined study areas. In addition to the demographic, lifestyle, traffic, and purchasing data most relevant to retail site location, these reports include data on workforce composition, concentration, and skills as well as crime frequency information. These reports would be more helpful to companies wishing to locate a production facility, service center, or sales office, or to a commercial real estate company wishing to present neighborhood information to clients.

You will use these resources in this chapter to assess the attractiveness of the site offered to Janice and Steven by the Twin Cities Redevelopment Task Force.

Log in to Business Analyst Online; select a geographic area for analysis

1. Load your browser and navigate to the Business Analyst Online site at
    `http://bao.esri.com`.

2. Enter the Username and Password that you selected when you subscribed to BAO in the appropriate text boxes. If you wish, view the Quick Start Video for a brief orientation to the Business Analyst Online system.

3. Click the Get Started button on the BAO home page. Then click the Select Geographies option, click the Metropolitan Areas (CBSAs) followed by Metropolitan Areas (CBSAs) by State. Click on Minnesota.

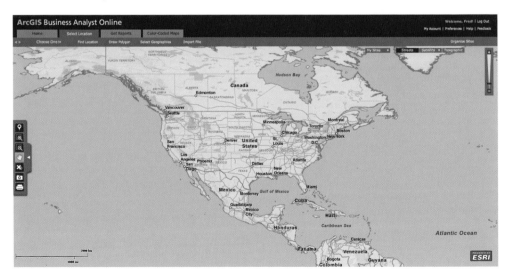

4. Scroll down to the Minneapolis-St. Paul-Bloomington, MN-WI CBSA and select it. Click Apply.

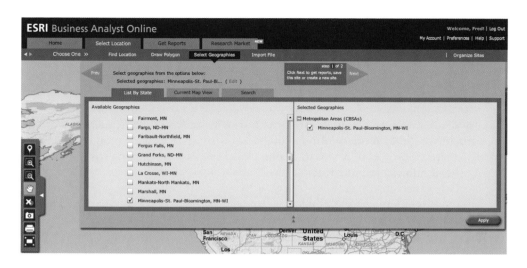

The map zooms to the target CBSA, highlights it, and displays its name in a dialog box. When your screen resembles the one below, close the dialog box.

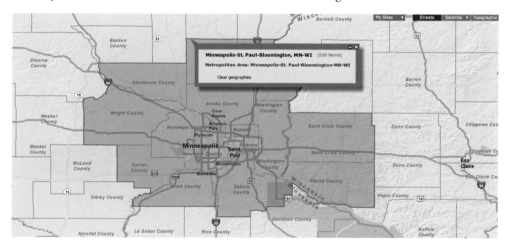

Create color-coded maps of income and home value

This study area matches the service area of the Twin Cities Redevelopment Task Force. Recall that the green-consumer segments Janice and Steven have chosen as their initial targets are distinguished by higher levels of income, education, and home value. You will use the color-coded mapping feature of Business Analyst Online to study the distribution of these characteristics in the Twin Cities area. Color-coded maps are also called thematic maps.

1.　Click the Research Market tab, then the Create Color-Coded Map command to open the settings pane for this function.

　　The command pane allows you to designate settings for the variable depicted in the map, the level of geography at which it will be presented, the colors scheme you wish to use, and the level of transparency of the map layer relative to the layers underneath it. By adjusting these settings, you will create thematic maps of income and home value at various levels of geography.

2.   In the Choose Variables drop-down box, review the categories of variables. Expand several categories to view the individual variables they contain. Expand CY[4] Household Income and select CY Median HH Income as the variable to display in the map. Click Geography, Auto Selection, then Counties.

Note that values for many variables are available as 1990 and 2000 Census figures, current-year estimates, and five-year projections. Thus you can use this tool to view chronological development of selected variables. Note as well that Median Household Income and Median Value of Owner Occupied Housing Units are available, but there is no single variable measuring educational attainment, which is captured in several different variables. This information will be obtained through the report function of Business Analyst Online.

3.   In the Colors selection box, select the Green monochrome option. Click the More Options button at the top right of the map window, then select the Quantile Method and 3 Classes.

The thematic map is generated and should resemble the one below, though the counties included and the boundary values of the classes might vary depending on the extent of your map. Review the contents of the map below. It displays the value of CY Median Household Income in the Minneapolis-St. Paul study area. In this map, each feature represents a county, though the geographic unit (block groups, census tracts, ZIP Codes, counties, states) will vary as you zoom in and out of the map. Recall that the median is the value that divides a set of features in half. For example, a median household income of $54,119 for a county means half the households in that county have incomes above that figure and half below.

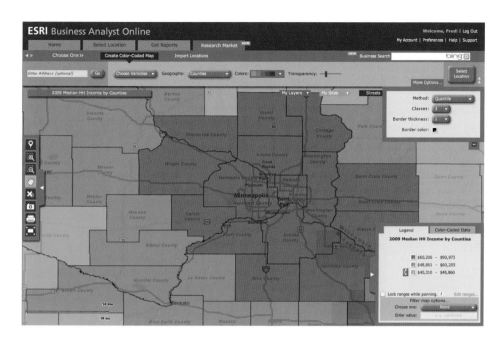

The Quantile method places the same number of geographic units into each of the classes designated. Thus your selections placed roughly one third of the features into each of three classes. This means that the median value for the features and about 16 percent of the features above and below this value are in the middle classification, while the top and bottom 33 percent of features are in the highest and lowest classification respectively.

4. To display the second variable of interest to Janice and Steven, expand the CY Home Value category and select CY Median Value: Owner HU in the Variable field, and Red monochrome in the Colors box. Retain the Quantile method and 3 Classes settings.

5. You may change the Geography level using the drop-down box to the left of the Colors box or by adjusting the Zoom bar at the top right of the map. Use one of these two methods to change the geographic level of the map to Census Tracts. Slide the Transparency bar from left to right to adjust the transparency of the layer.

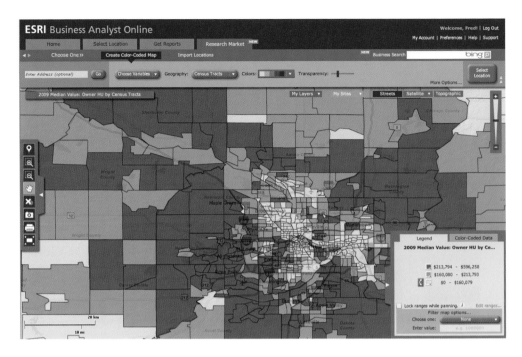

Observe the effect on the map, which should resemble the one above. The Legend reflects the new geography level and three classes in the data, each with approximately the same number of features as specified by the Quantile method. Scroll around the map to view the distribution of home values across the Minneapolis-St. Paul area. Note that, as you move the cursor around the map, the boundary of the geographic unit beneath it and the value of the variable being mapped are displayed in a pop-up window.

Use the Color-Coded Map command pane to adjust the map display. Try out different numbers of classes and the Equal Interval method. Adjust Transparency to view the street layers under the thematic map more or less clearly. Use different Geography levels to observe their sizes and distribution. Select different variables to view their distribution across the Twin Cities or to view changes over time within a single measure.

As you can see, thematic mapping is a powerful way to learn about the characteristics of a geographic region. You will use this tool later to select a site whose surrounding population would be attractive for LITGL's first store.

6. Click the Select Location tab to return to the original map.

Define alternative trade areas around a potential site

The Twin Cities Redevelopment Task Force offers tax incentives to entrepreneurs who revitalize existing facilities in the area. One such facility, at 2955 N. Second Street in Minneapolis, matches the criteria for the first LITGL store. Janice and Steven wish to evaluate the trade area served by this location to determine if it also matches the profile of their target consumers. You will use Business Analyst Online to define alternative trade areas for this site and select the most appropriate one for further study.

1. Click the Find Location tab and enter **955 N 2nd St, Minneapolis, MN 55411** in the Address field and **TCRTF Site** as the optional site name.

    Business Analyst Online geocodes the address and locates it on the map with a placemark and a pop-up window. You are ready to create trade areas around this point.

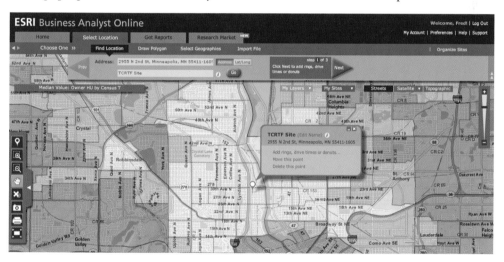

2. Close the pop-up window, then click the Next arrow at the right side of the Find Location command bar. Click the Rings tab, enter **1.0** miles as the first radius, **3.0** as the second, and delete the entry for the third radius, as you wish to use only two rings. Click Apply. In the Apply Rings window, confirm that the site is selected, then location, and click Apply.

    Notice the changes in the Business Analyst Online interface. The address you entered is assigned to a specific latitude/longitude location on the map (a process known as geocoding), the map zooms to that location, and two rings are drawn around the site at 1 and 3 mile radii respectively.

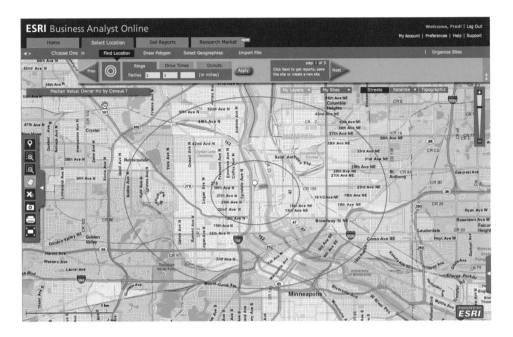

Examine this trade area for a moment. Note that it is bisected by an expressway running north and south, but serviced by smaller roads to the east and west. Further, more than a third of each ring lies across the Mississippi River, where access to the site would be limited to routes crossing the river. These factors suggest that the site might not be equally accessible to customers equidistant from it and, therefore, that the ring model might not be the most appropriate for this site.

Janice and Steven wish to evaluate a trade-area model that more accurately reflects the accessibility of potential customers to this site. You will create another study area that uses drive times to create an alternative trade-area definition.

3. Click the Drive Times tab, enter **3** minutes in the first drive-time box, **5** in the second, and delete the entry for the third, as you wish to use only two drive-time polygons. Click Apply. In the Apply Drive Times window, select the location and click Apply.

Business Analyst Online applies these changes and produces a new map with drive-time trade areas. Notice that these polygons reflect the limited access across the river, follow major highways for longer distances and minor roads for shorter distances.

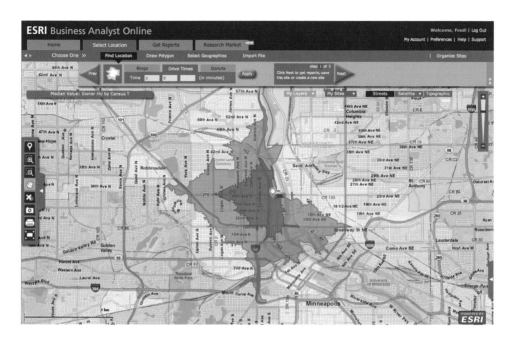

To return to the first map, click the Rings tab, then click Apply. To review the second map, click Drive Times, then click Apply. Toggle the two maps to compare the trade-area models. Note that the 5-minute drive time area extends to the 3-mile ring to the north and south along the expressway, but falls short of that ring to the east and west. Further, the trade area is larger on the west side of the river, reflecting more limited access on the east side resulting from the funneling of traffic over the river bridges.

Janice and Steven believe that this is the most appropriate trade-area model to use for comparative purposes. They now wish to select an alternative site based on a thematic map of income or home values. Before doing so, you will save the drive-time trade-area map for the first site.

4. With the drive-time trade areas displayed, click the Next arrow on the right of the Find Location command bar. Click the Save Site button, enter **TCRTF Site Drive Times** as the site name, then click Save.

Select alternative site and trade area

Recall that the thematic maps of income and home value in the Twin Cities area identified geographic areas with high values for these measures. For LITGL, these areas represent attractive prospective customers. Concentrations of these areas therefore, are favorable potential sites for the first LITGL store. You will recreate a thematic map and use it to identify an alternative store site.

1. Click the My Layers menu bar and select the 2009 Median HH Income map you created earlier to display it. Pan the map across the Minneapolis-St. Paul area. When you find concentrations of high income households, zoom into the map so data is displayed at the block group level.

2. Locate a concentration of block groups with high levels of CY Median Household Income. Adjust the Transparency bar and note the location of the site you wish to select. Click the Select Location tab, click the Pin icon in the toolbar on the left side of the map, then click a location within a cluster of high income block groups.

    This is the second site you will use for comparative purposes. Your map should resemble the one below, though your choice of location may be different.

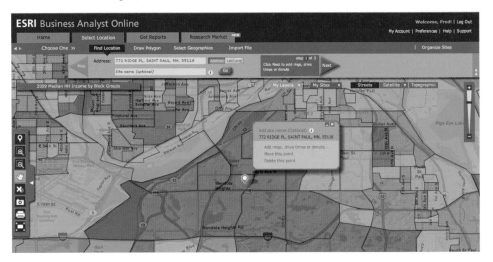

3. Click Next, select the Drive Times option, specify 3- and 5-minute drive times, and delete the entry in the third box. Click Apply. In the resulting dialog box, select both sites, then click Apply.

Business Analyst Online generates 3- and 5-minute drive-time polygons around both sites and displays them on the map with text boxes reporting their addresses. Your map should resemble the one on page 17, though the location of the second site and the shape of its trade areas might vary. Note in the example on page 17 that the elongated shape of the trade areas for the second site reflects its location on a major east-west road with small north-south streets. The shape of the trade areas for the site you selected will be similarly affected by its location relative to area roads and highways.

Compare sites with Business Analyst Online reports

Janice and Steven have identified alternative trade areas. One site is eligible for Task Force incentives and the other sits within a concentration of attractive customers. The next step is to compare the characteristics of these trade areas relative to their criteria for segment selection, income, home value, and education. They also wish to learn more about the consumer expenditure patterns and retailing environment in the two areas. You will use the robust reporting feature of Business Analyst Online to provide that information.

1. Click the Get Reports tab, then click Run Standard Reports to open the Report Table of Contents.

By default the menu lists all available reports. As new reports are developed and added to Business Analyst Online, the list on your screen may not match this list exactly. To view subsets of related reports, select the category you wish to review by using the drop-down menu in the View area. The most relevant reports for site-selection applications include the following:

Business reports include site maps and summaries of businesses by classification in selected study area. *Consumer Spending* reports include information on consumer spending in several categories. *Demographics reports* include information on population characteristics in selected study areas. Some focus on specific measures such as income, net worth, employment, age, and race, while others such as the *Executive Summary* and *Market Profile* reports provide general overviews of the demographic characteristics of the trade areas. *Maps* reports provide a collection of maps displaying the selected sites and their surroundings. *Tapestry Segmentation* reports provide information on the Tapestry Segmentation composition of the trade areas, while the remaining categories focus on detailed data such as traffic and site map information.

In addition, this screen displays the available trade-area options in a row across the top. The three entries there consist of the Minneapolis-St. Paul CBSA, selected at the beginning of this exercise, and the two sites you have identified. By running the same report for each of these three geographic units, you can compare the two sites directly to each other and both of them to the characteristics of the Minneapolis-St. Paul region as a whole. You may also generate quick site comparisons with the Create Comparison Reports option.

2. Click the Create Comparison Reports button to open the Comparison Report Setup window. Select the TCRTF Site and the second site for comparison. Do not select either of the two optional components. When your page resembles the one below, click the Housing button to generate the comparative report.

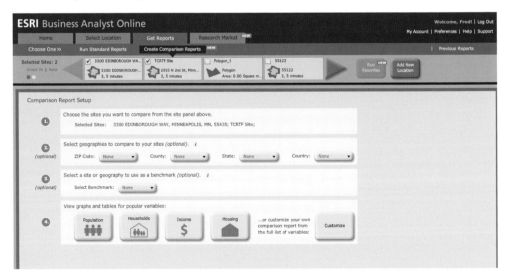

Business Analyst Online calculates the report data and presents it as a pie chart reflecting the distribution of owner-occupied, rental, and vacant housing units in the 5-minute drive-time trade areas around the two sites. Your screen should resemble the one below. Click on the Population, Household, and Income tabs to view those portions of the report.

The Comparison Report function provides an overview of site differences. For more detailed comparisons, you may use the Customize button on the Comparison Report Setup window to create tables and charts of variables of your choice. For additional comparative data for these sites, you will use the more comprehensive reporting functions of Business Analyst Online.

The real power of the Comparison Report is the ability to obtain a data dump. Click on the Create Custom button and then the table button to see a spreadsheet of your trade areas and their selected values. Save.

3. Click on the Market Profile Report to open its report description.

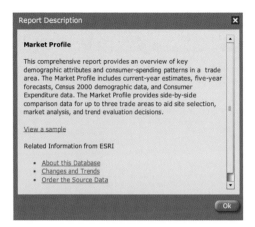

The report description feature serves several functions. First, it allows users to select more intelligently from a comprehensive collection of Business Analyst Online reports. In addition to the summary, the View a sample option allows users to view a sample of the report—its variables and layout—before ordering it. Second, it provides links to information about the database used to create the report as well as the most current demographic trends in this data. This information enhances users' understanding of the data in the report and, therefore, its credibility. A final link allows users to order source data directly if they wish to use a broader range of information from the data source.

4.   Click OK to close the Report Description box.

Take a few minutes to review other available reports. Although we will use a few reports to illustrate their value, you may also order additional reports to assess their value in the site-selection process. When you have finished your review, continue to the next step.

5.   Select all three sites in the bar above the Reports Table. Click the Add button to the left of the Market Profile report to add this report to the Selected Reports list at the bottom of the screen. Enter **Living in the Green Lane** in the Report subtitle box. When your screen resembles the one below, click Run Selected Reports. Business Analyst Online accepts the order, displays an Order Confirmation box and, when the reports are completed, a Reports Ready box. Click OK in each of the boxes.

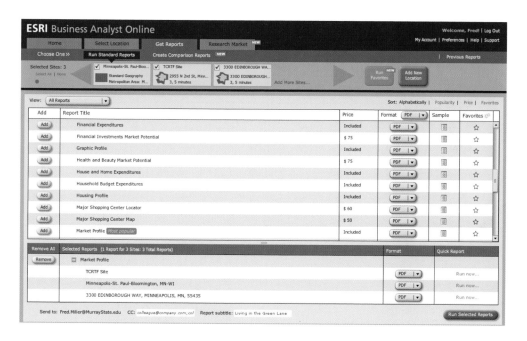

The Reports Ready box allows you to review the reports you ordered in PDF format, though you may also choose other report formats such as Excel. When you clicked OK in that box instead of opening the reports, Business Analyst Online stored them in your collection of Previous Reports. You will order several more reports, then open that list to explore the contents of the reports.

6. Unselect the Minneapolis-St. Paul CBSA in the bar above the Reports Table to limit your remaining reports to the two sites you have identified. Add the Business Summary, Executive Summary, House and Home Expenditures, Retail MarketPlace Profile, Site Map, and Tapestry Segmentation Area Profile reports to the Selected Reports list and Run them. Click OK in the Order Confirmation and Reports Ready boxes to store the new reports in your Previous Reports collection.

7. Click the Previous Reports button to open your collection of reports. Your screen should resemble the one on page 22, though the order of the report listing might vary. You may open any report as a PDF document by double-clicking its Report Name. You will use these reports to compare the two sites. Feel free, however, to order additional reports to orient yourself more fully to the depth of information available in Business Analyst Online Reports.

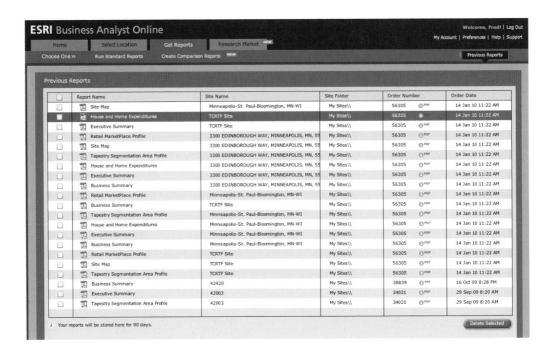

The reports in this list provide extensive information about the two sites. We have not included samples here, as the maps and data will differ based on your site selection. For general information, review the Executive Summary and the Site Maps. The maps display the environments of the two sites and their drive time market areas. The Executive Summary provides an overview of the demographic characteristics of the two market areas. Use this report to compare the two potential sites' characteristics relative to the demographic factors of greatest importance to Janice and Steven, specifically:

- Population and households currently and over time
- Household income currently and over time
- Home ownership currently and over time
- Median home value
- Educational levels, i.e., the percentage of adults over 25 holding an associate degree or higher

In addition to this general information, the reports in this list provide detailed data on population, consumer spending, and business characteristics of the two market areas. The Market Profile report expands the demographic overview in the Executive Summary into extensive breakdowns of population in the market areas by age, income, ethnicity, household size, type and value of housing, employment, and educational attainment. This report also identifies the top three Tapestry Segmentation[5] segments in each study area. Finally, this report provides consumer spending data for several broad categories of goods. This data includes spending in total and per household as well as a Spending Potential Index for each category. For example, in the HH Furnishings and Equipment category most relevant for LITGL, a Spending Potential Index of 120 means that households in the market area spend, on average, 20 percent more on these products than do U.S. households as a whole. Conversely, a figure of 80 indicates levels of spending that are 20 percent less than the U.S. average.

Information on population characteristics is also provided by the Tapestry Segmentation Area Profile report, which identifies the 20 most common Community Tapestry segments in the market areas and compares their concentration in the market areas to that of the overall population. It also reports similar comparisons for the Community Tapestry Life Mode and Urbanization groups in the market areas.

Business Analyst Online provides consumer spending pattern reports for several product categories. The one of most interest to Janice and Steven is the House and Home Expenditures report, which provides detailed spending information on various subcategories of this classification. It details overall spending, spending per household, and Spending Potential Indexes for each subcategory. LITGL's owners are most interested in the figures for Maintenance and Remodeling Services and Materials, but several other subcategories will be part of their planned product line as well. Finally, two reports in the list provide information on the business and competitive environment of the two market areas. The Business Summary report lists the number of enterprises and their employees by business classification in each of the market areas. For LITGL, the most relevant categories are Home Improvement in the Retail Trade Summary and the Building Materials category under Retail Trade. This report also provides information on daytime working, as opposed to residential, population for each market area, measured both in total employment and as a percentage of the residential population.

The Retail MarketPlace Profile report extends this perspective on the business environment by comparing total supply and demand for several types of consumer expenditures. Supply is reported as the total sales to consumers by stores in the area. Demand is total expenditures in the product category by residents of the area. Demand minus supply is reported as the Retail Gap and is used to calculate the Leakage/Surplus Factor. When this factor is positive, demand exceeds supply in the market area, a potential retail opportunity. This conclusion should be treated with caution however, as it might be reflective of zoning restrictions and/or concentrations of retail establishments in other areas which might be more attractive for LITGL due to their ability to draw retail trade. More extensive research on the competitive environment is necessary for understanding this situation properly.

Make a purchase decision on the TCRTF site

Janice and Steven must decide whether to purchase the TCRTF site for Living in the Green Lane's first store or continue the search for a more attractive site. Review the maps and reports you have created to recommend a course of action to them. Is the TCRTF site more or less attractive than the site you selected from your thematic map? Why? What factors distinguish the two sites and lead you to view one as superior to the other?

If you judge Site 2 to be superior and your recommendation is to forego this purchase, the site-selection decision becomes more complex, as TCRTF has no available property at that site. Therefore you must expand your search to include other potential sites at which facilities are available for purchase. You must also broaden your exploration of population characteristics, the business environment, and competitive situation to make a more knowledgeable selection. This will be your task in the next two chapters.

ROI considerations

At first blush, the return on investment calculation for this analysis appears unfavorable. Janice and Steven have decided not to purchase the TCRTF facility on the basis of their work with Business Analyst Online. However, the potential return on the information they have accumulated is considerable.

First, consider that this information is available at relatively modest cost. Business Analyst Online is available at annual subscriptions rates of under $3,000 and at considerably lower prices for terms as short as one day. While Janice, Steven, and you have spent considerable time working with the maps and reports in Business Analyst Online, this commitment is significantly more modest than that necessary to generate this information on your own.

On the other side of the ledger, the benefits of improving the site selection decision are considerable and, with the help of the Business Analyst Online report data, more quantifiable. Consider the following information drawn from reports similar to the ones you generated. This data covers the 5-minute drive-time market areas of the two sites.

Attribute	TCRTF Site	Site 2
Households	17,655	20,408
Household median income	$44,568	$82,426
% adults with associate degree or higher	29.8%	64.3%
% of housing owner occupied	48.8%	68.4%
Median value of owner-occupied housing	$109,717	$262,585
Expenditure per household	$1,646	$4,707
Total expenditures	$29,060,130	$96,072,902
Incremental revenue per share point (equal SoM)		$670,128
Incremental revenue per each additional share point		$960,729

Table I.2 Business Analyst Online Site Information: Building supply and home improvement retail purchases.

The first row is the number of households in the 5-minute drive time of each market area. The next four rows report the attributes Janice and Steven are using to define attractive prospective customers based on their understanding of green-consumer segments. These values indicate that Site 2 is significantly more attractive than the TCRTF. This conclusion is supported by the higher levels of maintenance and remodeling services and materials expenditures, both in total and per household, in Site 2's market area.

Based on these considerations, Site 2 is the more attractive market area, but what is the financial benefit of choosing Site 2 over the TCRTF site? The final two rows contain data relevant to that question. The first reports incremental revenue in Site 2 versus the TCRTF Site for each point of market share. The value of $670,128 in the Site 2 column means that a 1 percent market share in this area produces $670,128 more than a 1 percent share of the TCRTF Site market area. Thus, if LITGL could achieve a 5 percent share of market in either site, its revenue would be $3,350,640,170, or 5 times $670,128—higher at Site 2 than at the TCRTF Site.

Further, Site 2's population is clearly a better match with the green-consumer profile than is the population of the TCRTF Site. Thus, it is quite likely that LITGL could achieve a higher share of the market in Site 2's market area than it could in the TCRTF Site's market area. The final row in the table indicates that each point of such additional market share would increase LITGL's revenue by $960,729. Thus, if the company achieves an 8 percent share of market in the Site 2 market area versus a 5 percent share in the TCRTF Site's market area, it would add an additional $2,882,187 (3 times $960,729) in revenue, producing total incremental annual revenue of $6,232,827. Finally, note that these figures reflect only two categories of House and Home Expenditures from that report. If, as it plans, Living in the Green Lane offers additional products and services to its customers, these incremental revenue figures would be more substantial yet.

As these estimates indicate, the potential benefits of bringing more comprehensive, reliable information to bear in the site selection decision are substantial. This is especially true in relation to the modest costs of using Business Analyst Online to support this decision-making process.

Summary of learning

In short, you have learned in this chapter how to use Business Analyst Online to create a study area, to display data attributes on a map, to locate potential sites in two ways, to create market areas around those sites in three different ways, to order relevant reports for use in comparing alternative sites, and to use those reports in identifying attractive sites.

In the following chapters, you will expand this site selection process by using the broader range of tools and expanded environmental scanning capabilities of ESRI Business Analyst Desktop 9.3.1.

Notes

1. This is a fictional organization. Any resemblance to an actual organization is coincidental. However, economic development organizations frequently use business GIS tools, including ESRI Business Analyst, in their efforts to support existing companies and nurture new businesses. Thus it would not be unusual that an entrepreneur's first encounter with these technologies would be in this context. An example would be the South Bend Small Business Development Center. `http://www.esri.com/library/fliers/pdfs/CS-southbend.pdf`.

2. Schaffer, Paul. 2007. New study: Americans reach environmental turning point, companies need to catch up. *Environmental News Network*, August 22. `http://www.enn.com/top_stories/article/22186/print`.

3. Kannan, Shyam. 2007. Unveiling the green homebuyer. *Urban Land*. June, pp. 106–9.

4. Throughout this book we will refer to data as CY or FY values. CY or "current year" designates the estimates of values for the most recent year available. FY or "future year" refers to projections of values five years beyond the current year. As ESRI updates Business Analyst datasets, you might be working with more recent data than that depicted in the graphics in this book. Thus, the CY/FY convention will place that data in the proper context and avoid designating specific dates that don't match your data.

5. Tapestry Segmentation Segments are classifications of U.S. neighborhoods into 65 distinct lifestyle clusters based on factors such as age, income, education, housing, purchasing patterns, and values. This system will be used extensively in segmentation analysis in later chapters. It serves here as another means of comparing the two market areas under consideration.

Part II

Business environment analysis with ESRI Business Analyst Desktop

Relevance	Entrepreneurial new businesses need to develop responsive marketing plans to support success, and convincing business plans to secure funding.
Business scenario	LITGL must demonstrate to potential investors its understanding of the business environment in which it will operate. This includes identifying concentrations of potential customers within the context of the retail environment, including retail centers, competitors, and transportation infrastructure.
Analysis required	LITGL must extend its understanding of target customers, seek out concentrations of them, and display this information on maps that also symbolize information on competitors and retail shopping centers.
Role of business GIS in analysis	Identify concentrations of favorable customer characteristics with thematic mapping. Gather information on competitors and retail shopping centers and symbolize this information in layers of thematic maps.
Integrated business GIS tool	ESRI Business Analyst Desktop.
ROI considerations: Cost of business GIS	Business Analyst Desktop purchase, training of researcher.
ROI considerations: Benefits of business GIS	Increased sales from favorable location decision. Potentially lower loan costs if lender views the project as less risky based on customer data.
Environmental impact of business decision	Increased efficiency of serving customers due to proximity. Lowering of customer fuel costs by convenient location within transportation infrastructure.

Table II.1 Executive summary

The Living in the Green Lane scenario

While their Twin Cities Redevelopment Task Force study did not result in a final site decision, it did introduce Janice Brown and Steven Bent to the process of market area analysis. In addition, their work with ESRI Business Analyst Online in that process convinced them that integrated business GIS is an indispensable tool in the planning process. As a result, they offer to promote you from GIS planner to Living in the Green Lane's business GIS analyst. When they agree to your request to purchase a license for ESRI Business Analyst Desktop to support this position, you accept the offer.

Your initial assignment is to help Janice and Steven complete LITGL's business plan. The key goal of that plan is to convince investors that the LITGL concept will be successful in the Twin Cities area. It must describe the company's business concept and its attractiveness to target customers. It must include analysis of the potential customer base as well as a detailed description of the retail environment in the area. The latter includes the location and size of retail shopping centers that attract shoppers as well as information about competing stores that might siphon sales from LITGL's location. It also involves the transportation infrastructure of the Twin Cities area relative to concentrations of households in the green-consumer profile. Finally, it must identify the site of the first LITGL store and explain the justification for selecting that site.

In the next chapter you will assess available sites and select the most attractive one for the first Living in the Green Lane store. That decision, however, must be grounded in a thorough understanding of the company's consumer and competitive environments. Achieving that understanding is the focus of this chapter. To complete it, you will perform four tasks with Business Analyst Desktop.

First, you will open a Business Analyst project and create a Business Analyst study area to organize your work. This will establish the geographic focus of your analysis of the Twin Cities area and display the transportation infrastructure of the region.

Second, you will use thematic mapping to explore the distribution of population characteristics across the Twin Cities area. In doing so, you will be extending the procedures that you applied in your population analysis in chapter 1. Specifically, you will incorporate more extensive data attributes, classification schemes, and symbolization options into your work. However, you will continue to concentrate on the population characteristics of the green-consumer profile, as this is the customer segment upon which the success of this enterprise depends.

Third, you will symbolize important features of the competitive retail environment by extracting appropriate business information from Business Analyst datasets and symbolizing them on your map. You will include both retail centers that attract customer traffic and competing home centers that are Living in the Green Lane's most direct competitors.

Fourth, you will integrate all these resources, as well as satellite imagery from ArcGIS Online, into a common mapping framework and use Business Analyst tools to discover potentially favorable sites in the Twin Cities area and create a market area report on one of those sites.

Integrated business GIS tools in consumer and competitive environmental analysis

One of the core functions of integrated business GIS is to help entrepreneurs understand their business environments more fully and communicate that understanding to internal and external stakeholders more effectively. This environmental scanning process is the essential starting point for strategic marketing management and the foundation for well-crafted business plans. Integrated business GIS streamlines that process considerably by accumulating various types of business information, integrating it into map documents and revealing the spatial relationships between discrete features of the business environment. The resulting maps display multiple environmental factors simultaneously, symbolize the most relevant characteristics of each type of data visually, and communicate the spatial relationships of factors clearly.

This process builds on the most basic function of mapmaking: symbolizing important features of the physical world and their spatial interrelationships. In this case, symbology will focus on major freeways in the area, population characteristics of the green-customer profile, the location and size of retail shopping centers in the area, and the location and size of competing home centers.

As the core function of a business is to identify and serve customers, the first step will focus on population characteristics. Thematic mapping, which you performed in chapter 1, is the most powerful GIS tool for this purpose. Recall that thematic maps display the geographic distribution of selected attributes across a region of interest. You will use it to help Janice and Steven understand their market environment more fully.

Thematic mapping decisions

Within the rich symbology and data environments of integrated business GIS systems, effective thematic mapping involves a host of mapping decisions, ranging from the data to include in the map to the symbols and classification schemes used to display it. Let's consider each in turn.

What data to symbolize

Integrated business GIS systems offer incredibly rich datasets. In the desktop environment, for example, Business Analyst includes data on more than 1,500 attributes at the U.S., state, county, CBSA, Designated Market Area (DMA), ZIP Code, census tract, and census block group levels. Identifying the most relevant attributes to use in business decisions can be a daunting task and can vary significantly between different types of business problems.

To simplify this decision, Business Analyst Desktop organizes attributes into groups of similar data. The most relevant groups for marketing and retail site selection decisions include the following:

General summary and population demographic data
This group includes overview attributes of households including age, housing size and status, income and home ownership characteristics, all in current-year estimates and five-year projections. This group also includes more detailed data on age, gender, and ethnic distribution as well as employment by industry and occupation. This group has broad relevance for marketing, retail site, and commercial site selection as well as general population studies.

Financial resources data

This group includes more detailed attributes relative to family and household income, household disposable income, net worth, and home value, including information about distribution of households across ranges of financial resource data. This information is relevant to a wide range of marketing decisions. In addition, this group includes an extensive breakdown of households by income for each of 14 different age groups. This data is of great value to companies that segment customer markets by age and income. This data also includes current-year estimates and five-year projections.

Tapestry segmentation data

This group includes data on the distribution of both households and adults across 65 Tapestry Segmentation lifestyle segments, 12 Life Mode groups, and 11 Urbanization groups. This data is vital to marketers using lifestyle segmentation by both defining lifestyle segments and, as we shall see later, tying those segments to very specific purchasing, media exposure, activities, and values measures.

Consumer expenditure data

This group of data details expenditures on a wide range of product and service categories. This data is very useful in displaying purchasing patterns in aggregate and on a per-household basis.

2000 and 1990 census data

This group includes selected attributes from the 2000 and 1990 censuses. It is useful for two purposes. First, it supports more extensive chronological comparison of important population measures. Second, it includes data from the 2000 census on factors such as household composition, travel-to-work methods and travel time, levels of rent/mortgage payments, and educational attainment, among others. Although more dated than the current-year estimates, this data is of value to companies for which these factors are important market indicators. Vehicle ownership and driving times to work are examples of this type of data.

Within this rich range of possibilities, the objective of thematic mapping is to display the attributes most relevant to the current marketing decision. Janice and Steven have identified household income, home value, and educational attainment as distinguishing profile characteristics of green-building customers, so these will be the attributes you include in the thematic mapping process.

Of these attributes, home value is the most straightforward, with both median and average values available at each level of geography. For reasons discussed below, you will symbolize median values.

Household income is more complex in that several attributes reflect the financial position of households. Figures on per capita income, family income (average and median), household income (average and median), and disposable income (average and median) are available. Household net worth (average and median) would also be worthy of consideration as a potential source of funding for substantial building projects. Janice and Steven wish to focus on household income for three reasons. First, this is the measure in the green-customer profiling studies and, therefore, most appropriate for the screening process. Second, there is no reason to limit LITGL's target market to family households rather than all households. Third, although net worth is a relevant factor for homeowners considering large building projects, Janice and Steven anticipate that the largest portion of sales will be for more modest projects, for which income is the more relevant measure of available financial resources.

The final decision for income data calls for determining whether to work with median or average values for household income. Median household income is the value that bisects the households in the selected geographic area—half earning more and half earning less. Average household income is total household income in the area divided by the number of households. However, as a measure of the financial resources available to the households in the area, average household income can be misleading. Typically, average values are greater than median values. This is true for the Twin Cities CBSA overall and in 98.8 percent of its 2,241 block groups. For the CBSA, average income is 27 percent higher than median income. This difference is even higher for 26 percent of the area's block groups.

The effect of this distribution is illustrated in the chart below, which displays the distribution of households in the Twin Cities CBSA by levels of household income. The households represented by the beige bars on the left fall below both the median and average values for household income. The brown bar represents households in the $60,000–$75,000 range and contains the median value of $73,045. The red bar represents households in the $75,000–$100,000 range and includes the average value of $91,608. The bars to the right in lighter red colors represent households with incomes above both the average and median values.

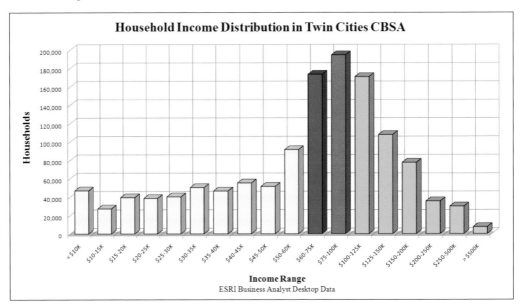

Figure II.1 Distribution of households in the Twin Cities CBSA by levels of household income

If households in the $75,000–$90,000 interval are evenly distributed across that range, then only 37 percent of households in the CBSA have income at or above the region's average value of $90,764. By contrast, 50 percent of households have incomes at or above the median value of $71,394. Thus, median household income reflects the level of income available to a higher percentage of area households than does average or mean household income. As LITGL wishes to target as many households as possible in the higher income market areas it serves, median household income is the more appropriate measure for this study.

What symbology to use

Once data is selected, the next question is how to symbolize the most germane attributes on a map. For point features, such as stores, shopping centers, competitors, and customers, the options include graphic symbols, their relative sizes, and their colors. For example, competitors can be represented with one symbol, shopping centers by another, with the size of each symbol reflecting the annual sales of each entity.

For polygon features such as the various geographic units for which data is available, the options include a variety of color and/or hatching schemes and graphs within each unit, or dots that represent the concentration of an attribute in each unit. Of these, dot density is the most straightforward and is ideal for depicting concentrations of relevant attributes across a geographic region.

Graphs are useful for depicting values of similarly scaled attributes within each geographic unit and for displaying several attributes simultaneously. For example, a bar chart in each of several ZIP Codes can depict several percentage values (home ownership, educational attainment, single-family households) in a single bar graph. A pie chart similarly configured can depict population characteristics by factors such as gender, ethnicity, and age. When set against a transparent background above another data layer, graphs increase the information presented in the map and, potentially, its communication value.

Color and/or hatching schemes display each geographic unit in a color or pattern, which reflects its value for a relevant attribute. All units with values for the attribute within an assigned range are displayed with a common color or hatching pattern. By using a combination of color, hatching, and transparency settings, users can depict multiple attributes for the same unit simultaneously. Color schemes and, to a lesser extent, hatching schemes, are useful for displaying a wide range of population attributes relevant for marketers. Indeed, thematic maps based on color schemes typically provide the geographic context over which other data layers are symbolized. Thus, such vital factors as the distribution of stores and competitors relative to concentrations of target customers become quickly and visually discernible.

What classification scheme to use

A classification scheme involves two factors: The number of classes to create and the method for assigning features to each class based on attribute values. Consumers of map documents interpret geographic units displayed in the same color as being similar to each other but different from units displayed in other colors. The classification method employed determines the degree of those similarities and differences as well as the basis upon which they are determined. Classification method options include the following:

Data-driven methods
These methods create classifications based on the distribution patterns of the data being displayed. For example, the standard deviation method is based on standard measures of central tendency and dispersion within a data distribution. It uses the mean and standard deviation of a set of values to assign features to classes based on the number of standard deviations their attribute values lie above or below the mean of the dataset.

The Jenks method, also called Natural Breaks, identifies natural clusters within a dataset by maximizing statistical variance between groups while minimizing it within groups. While useful for revealing patterns of natural clustering, this method is influenced by outliers and can produce classes containing very uneven numbers of features.

The geometric interval method creates a series of intervals based on a geometric coefficient and its inverse. These values are calculated to minimize the sum of squares within each class. This method is designed to display continuous data effectively.

Interval methods
In the equal interval method, the user defines the number of classes to be displayed on the map and ArcGIS uses the range of values for the selected attribute to create a series of ranges of equal size. The defined interval method uses a reverse process; the user designates the size of each range, say $40,000 for an average home value attribute, and ArcGIS creates the number of classes necessary to depict the full dataset in ranges of $40,000 each. These approaches are useful in projects focusing on the distribution of features across a predetermined range of values.

Quantiles
In the quantile method, features are ranked on a relevant attribute, then grouped into classes containing approximately equal numbers of features in each class. In applying this method, the user defines the number of classes to be used. So, for example, quartiles use four classes and quintiles five. Since the classes are not directly derived from the distribution of attribute values, features with very similar values are often placed in different classes.

On the other hand, the distribution of units relative to the median value of the relevant attribute is clear. A three-quantile scheme creates easily understood high, medium, and low classes with the median in the medium class. In a quartile scheme, the median separates the top two classes from the bottom two. In a quintile scheme, the median lies within the middle of the five classes displayed. You used a three-quantile scheme in chapter 1 and will apply quartile and quintile schemes below.

Manual creation
As the name suggests, this method allows the user to identify class boundaries directly. It is useful when a company has a clear definition of its target consumers, for example, and wishes to match the classification scheme with values from that definition.

What color combinations to use[1]

The thematic mapping tools within Business Analyst provide a wide range of color and hatching options. Users can select from a multitude of preset color schemes, design their own, or customize color and hatching patterns individually for each class of features. Proper use of these tools requires designing color schemes to support the communication objectives of the thematic map. Relevant considerations include the following:

Type of color scheme

Users may select schemes that use variations of a single color, gradual transition between two colors, or distinctive colors for each class. The first approach, known as sequential schemes, is most appropriate to communicate progressive values of an attribute—such as household size—within the area of interest.

Gradual transition schemes, known as diverging schemes, communicate feature values relative to a significant value somewhere in the central range of a distribution. For example, a map might display values at increments of the standard deviation above the distribution mean with variations of one color, and those below with variations of a second color.

The choice of a sequential or diverging scheme depends upon the communication objective of the map. If the goal is to display variations across the full range of the attribute, a sequential scheme is more appropriate. If the goal is to display variation from a central measure, such as average or median, a diverging scheme is more appropriate. This approach would be useful in a quartile classification scheme to distinguish those features below the median value—for, say, home ownership in the Twin Cities CBSA—from those above the median value.

The distinct color scheme, known as a qualitative scheme, is useful for communicating segment designations. For example, a bank offering different services to households at different levels of net worth would assign features to different segments based on that value. A map depicting these segment assignments with distinct colors for each would be most useful in communicating the geographic distribution of each segment.

Relevance of colors to attributes

Maps can communicate more effectively if there is a perceived relationship between the color schemes they use and the associated attributes they represent. Land-use maps, for example, commonly depict arid regions with shades of brown and forested regions with shades of green. Similarly, ocean maps depict increasing depth with darker shades of blue. Red and blue map colors have become so associated with voting patterns in the United States that the terms "red state" and "blue state" are understood perfectly when used outside the context of any map. Imagine how confusing a map would be that reversed any of these patterns.

The most obvious use of this approach in business GIS is selecting green to depict attributes expressed in currency. This might include a variety of income measures, home value, net worth, and expenditure patterns. Expenditures may also be depicted in colors appropriate to the spending category, e.g., swimming pool ownership in shades of blue or wine consumption in shades of burgundy. Similarly, green-to-brown scales might also be used to indicate segment membership within the green-customer profiling schemes used in this project. A green/amber/red scheme is useful for distinguishing primary segment targets from secondary targets, and both from segments that will not be targeted.

Balance of colors
This refers to the relationship between colors in the scheme. Darker or more vibrant colors create a stronger visual impression than do more subdued hues, creating the impression that the features they represent are larger or more important than they actually are. Likewise, strongly contrasting colors can create the impression of more significant differences in the values of the attributes they represent than is really the case.

A second consideration here is the range of colors to use relative to the classification scheme selected. If several classes are being displayed, a scheme using only shades of a single color may render it difficult to discern the exact class to which a feature belongs, and even harder to determine if two features in different parts of the map are in the same class or not. This difficulty can be addressed by limiting the number of classes or using a more diverse color scheme.

Role of the thematic layer in the map
The selection of a color scheme for a thematic map must support other layers in the map as well. If the primary purpose of the map is to display the distribution of attribute values in the area of interest, the thematic layer may well be the most important and visually dominant layer in the map. Often, however, thematic maps serve the dual purpose of communicating a distribution while also providing a context within which other meaningful layers are displayed.

In chapter 1, for example, you used a thematic map as a backdrop for the market area polygons for the two sites you considered. The maps you design below will identify concentrations of attractive population characteristics as well as provide a context for displaying the locations of shopping centers and competing home centers. In this setting, the thematic map must use a color scheme which meets its communication objective; but it also provides a background in which these additional point feature layers are visible and their attributes clearly distinguishable.

Symbolizing point features
The classification schemes described above are available for point features such as stores, shopping centers, competitors, and customers. However, they are displayed differently. Options for point features include the use of symbols, their size, and their color. For example, sales figures for stores or shopping centers can be symbolized by using circles of the same size with different colors for each class, or circles of the same color whose size reflects the class of which they are a member.

Categorical characteristics such as hardware stores versus appliance stores also may be expressed by using differing symbols for each category or the same symbol with variations in color for each category. Further, by symbolizing different types of stores in different map layers, these options may be combined to display sales levels as a range of sizes or colors by, for example, using different symbols for each store layer to distinguish them. Thus hardware stores might be symbolized with red circles (or, more creatively, hammers) and appliance stores with blue triangles (or toasters). In each case, graduated sizes of the relevant shape could indicate annual sales levels as well.

Business Analyst Desktop provides significant information on shopping centers and businesses, including estimates of sales, size, number of employees, and, for shopping centers, renovation information as well as the identity and size of major anchor stores. If additional data is necessary for analysis of the business environment, users also may add their own enterprise or third-party data as needed.

Using Business Analyst Desktop to research consumer and competitive environments

Business Analyst Desktop 9.3.1 offers unique capabilities for exploring the business environment of an enterprise. Its rich symbology options, extensive flexibility in defining symbology, and extensive data collections provide powerful capabilities to users. These capabilities are expanded through the ability to integrate visual data on several different components of the business environment simultaneously, and to do so in a medium that inherently captures the spatial relationships of those components as well. The result is a powerful tool for understanding threats and opportunities more clearly and communicating them more effectively. You will apply these tools to the task of analyzing LITGL's business environment to identify the opportunities the company should exploit in its business plan.

Chapter 2

Thematic mapping with ESRI Business Analyst wizards and layer properties

ESRI Business Analyst provides a range of tools to facilitate thematic mapping. The tool includes basic symbology options for point, line, and polygon features. The settings and classification schemes available in Layer Properties functions extends these capabilities significantly, providing substantial support for customizing map content. In this chapter, you will learn how to use these tools to perform fundamental thematic mapping tasks.

Run Business Analyst Desktop; open a project and create a study area

1. Click Start, Programs, ArcGIS, Business Analyst, BusinessAnalyst.mxd to run ArcMap, load the Business Analyst Extension, and then load the default Business Analyst map.

2. Click OK when the Business Analyst Map Projection dialog box appears, then close the Business Analyst Assistant window on the right of the screen.

   Your screen will resemble this default Business Analyst map. It displays a map of the United States and several of the country's major cities. Notice in the Table of Contents on the left that the map has two group layers, named Business Reference Layers and Map Layers. Map Layers includes several additional group layers, each composed of several layers whose visibility depends upon the scale and the extent of the map. You may toggle the display of all the layers in a group layer by selecting and unselecting the group layer title. In addition, you may expand or contract the group layers to control how much of the Table of Contents is visible. The same is true of individual layers. For example, turn on and expand Demographic Layers. Then expand the State Areas layer to view the Legend for the current map, which indicates that it is displaying state values for Median Household Income.

   You will organize your workspace by using an existing project to contain the files you will create in this analysis.

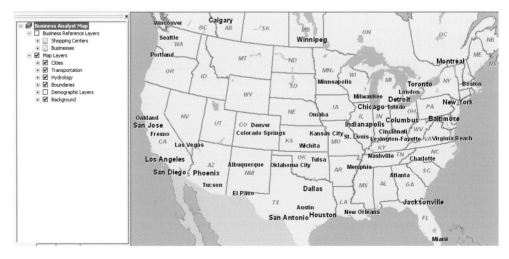

3.   Locate the Business Analyst toolbar, click the dropdown box on the left of the toolbar, click Active Project, then click the LITGL Minneapolis St Paul project to select it. This project was created when you installed the data for this book. Typically, you would create a new project of your own to begin a new analysis.

The project file for your work is ready. Within it, you will now define a study area for your analysis, in this case the Minneapolis, St. Paul, Bloomington Core Based Statistical Area (CBSA).

4.   Click the drop-down box on the Business Analyst toolbar and click Study Area to open the Study Area wizard. In the first screen, select Create New Study Area, and click Next. Select By CBSA from the available options, select From a list, click Next. Select Minneapolis-St Paul-Bloomington, MN-WI from the options, move it to the right box with the arrow key, and click Next. Enter **LITGLMinnStPaul** in the Name box and an appropriate description in the Comments box. Click Finish.

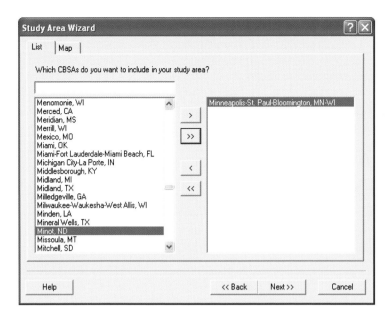

Several things happen. The map zooms to the study area you have created, which is outlined in a bold border. It is also added as a new group layer to the Table of Contents. Within the study area, major freeways and the boundaries of Minneapolis and St. Paul are displayed. Look at the Table of Contents to see why. Expand Freeways under Transportation in the Map Layers group layer. Note the difference in the selection boxes for the State View and Interstate layers. The lighter color and small bottom bar of the Interstates and Streets (Regional) layers indicate that they are available, but not displayed at all scales. This scale dependency, which you can customize in a layer's properties, means that different layers will display as you zoom in and out of the map. These settings ensure that the more detailed layers appear only on larger scaled maps, where they are more discernible and useful.

You have created a project and study area for this study. You will now clip the data layers in the Table of Contents to the shape of the study area. This will make your thematic maps easier to read and concentrates users' attention on the area of interest.

5. Right-click anywhere in the map area, click Data Frame Properties. Click the Data Frame tab. In the bottom half of the dialog box under Clip to Shape, select Enable then click the Specify Shape button. In the Data Frame Clipping dialog box below the Data Frame Properties box, select Outline of Features, then select LITGLMinnStPaul in the Layer box. Click OK, then Apply.

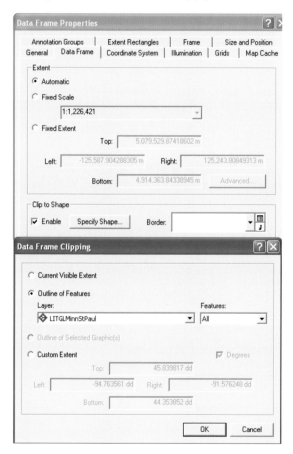

All the features have been clipped to the boundaries of the study area, which appears against a blue background. You will change the background color to white.

6. In the Data Frame Properties box, click the Frame tab and about halfway down, click the drop-down arrow to the right of the Background box and set the background color to white. Click OK to apply the setting and close the box.

Your map should resemble this one. You are now ready to begin symbolizing population, retail, and competitive information within the study area.

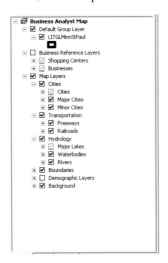

You will now save this map document so you can return to it easily.

7.   Click File, then Save As. Navigate to the C:\My Output Data\Projects\LITGL Minneapolis St Paul\CustomData\ChapterFiles\Chapter2 folder where you wish to save the map document. Enter **LITGLBusinessPlan1.mxd** in the File Name field, click Save.

This file name will now appear as an option when you run Business Analyst, allowing you to return directly to this map.

Create a chart map of home ownership with the Thematic Mapping wizard

Business Analyst Desktop offers several approaches to thematic mapping. You will begin by using the Thematic Mapping wizard from the Business Analyst toolbar to display home ownership rates at the county level within the study area.

1.   Expand the drop-down menu of the Business Analyst toolbar, click Tools, then Thematic Mapping to open the Thematic Mapping wizard.

2.   In the drop-down box, select County Areas (not County Boundaries) as the layer you wish to symbolize. Click Next.

3.   In the next window, select the Show a chart option and select Pie Chart from the drop-down box. Click Next.

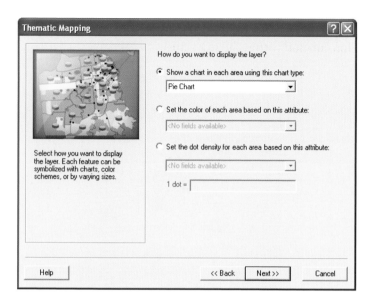

4. Select the attributes Current Year (CY) Owner Occupied HUs, CY Renter Occupied HUs, and CY Vacant Housing Units in the box on the left side of the window. Click the single arrow button to move these attributes to the right box to include them in each pie chart. Click Next.

Note: The attribute names here use CY for current-year data and FY for future-year projections. As Business Analyst provides annual updates for its demographic and Tapestry Segmentation datasets, the data names listed in the images in this book might include different years from those you encounter in the software. The CY/FY convention, therefore, will allow you to use the appropriate data attributes in spite of this variation.

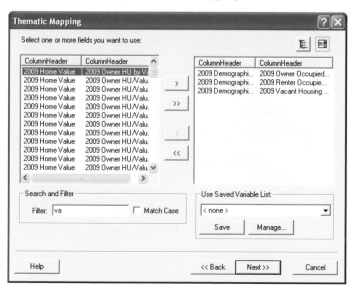

5. Select a color scheme from the available options. Click Finish. The wizard will process your settings and produce a map of the counties in the Twin Cities CBSA. Each county will contain a pie chart displaying the portion of housing units occupied by owners and by renters as well as vacant housing units. The map should resemble the one below, though the colors might vary. If this map doesn't display immediately, confirm that the Demographic Layers group layer is turned on and that only the County Areas layer is turned on. If pie charts still do not appear, also make sure you have a U.S. or at least a Minnesota license for the software.

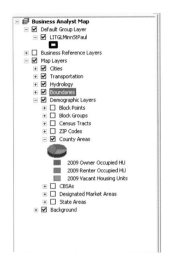 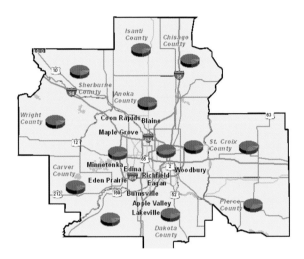

Data for each county is displayed by pie charts, but several of these pie charts are hidden behind other features and county boundaries and are not clear. You will turn several layers off and adjust the properties of the County Areas layer to correct this.

6. In the Table of Contents, uncheck the boxes to the left of Transportation and Hydrology layers to turn them off and make the map less cluttered.

7. Double-click the County Areas layer to open its Layer Properties window, which will display the Symbology tab. Click the colored box just to the right of the Background field to open the Symbol Selector window.

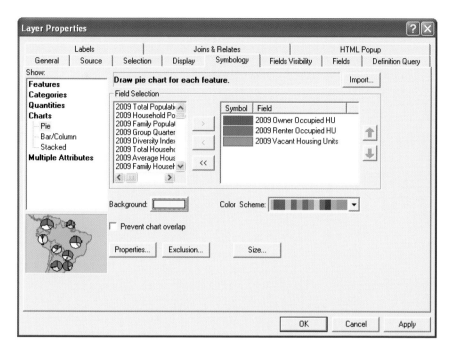

8. In the Symbol Selector Window, set the Outline Width to 2 pixels and the Outline Color to Black. The window should resemble the one below. When it does, click OK.

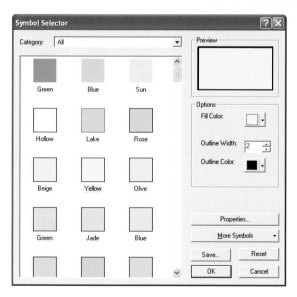

The map will now resemble the one on next page, though the colors may vary.

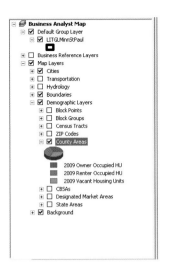
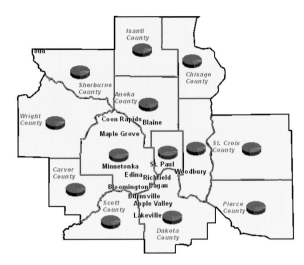

Review the charts for each county. For Living in the Green Lane, homeowners, who are more likely to invest in major projects than are renters, are primary target consumers. Renters are a secondary market with interest in making smaller investments in energy-saving technology to lower utility costs. While there are variations in the percentage of owned homes across the area, this map indicates that the overall level of home ownership is high, a favorable market indicator for LITGL.

Create a map of median home value with the Thematic Mapping wizard

Home value is another dimension of the green-consumer profile identified by Janice and Steven. Owners of high-value homes are more likely to invest in environmentally friendly renovations and to have the financial resources of income and home equity lines of credit to allow them to do so. You will create a thematic map of median home value to display the distribution of this attribute within the study area.

1.   Uncheck the County Areas layer to turn it off. Expand the drop-down menu of the Business Analyst toolbar, click Tools, then Thematic Mapping to open the Thematic Mapping wizard.

2.   In the drop-down box, select Block Groups as the layer you wish to symbolize. Click Next.

3.   In the next window, select the Set the color of each area based on this attribute option. Select CY Median Value: Owner HU as the attribute to display. You will find it in the CY Home Value group, roughly halfway down the list of attributes. Your screen should resemble the Thematic Mapping window at the top of page 49. Click Next.

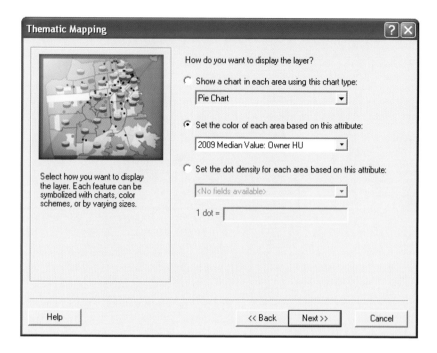

4.   In the next window, select a color scheme in a subdued color with clear contrast as the color scheme. Select the Natural Breaks as the classification scheme. Do not change the default normalization option, which is *none*, as this attribute is already at the household level.

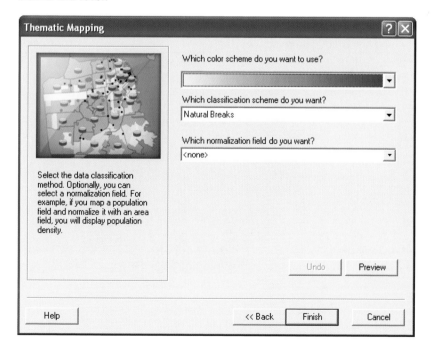

5. Click Finish to produce the thematic map. In the Table of Contents, check the Block Groups box to view it, and, if necessary, expand this layer to view the legend. Your map should resemble the one below. If it does not, confirm that the Block Groups layer is turned on and the County Areas layer is turned off.

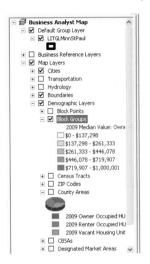
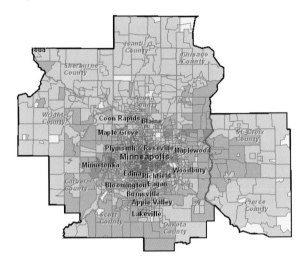

Observe that there are relatively few block groups in the classes with the highest values. Remember that the Natural Breaks classification scheme can produce this result. To create a map with more balance between the number of block groups in each class, repeat steps 1–5, this time selecting the Quantile classification scheme rather than Natural Breaks. You may also try another color scheme, if you wish. The resulting map will resemble the one below, though your color scheme might differ.

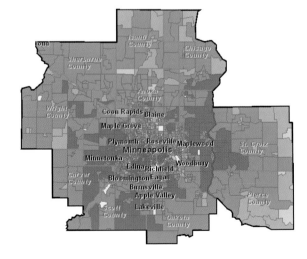

In this map, the number of block groups in each class seems more balanced, with the national median value of the block group numbers falling into the middle class. Check the box to the left of the County Areas layer to turn it on. The pie charts you created in that layer now appear on top of the new map. This allows you to explore home ownership and home value patterns simultaneously. As you may wish to return to this point in your analysis, you will set a bookmark here.

6.   Click Bookmarks on the menu bar, then Create to open the Spatial Bookmark dialog box. Enter **HomeOwnValueMap** as the bookmark name. Click OK. Click Bookmarks again to confirm that this bookmark appears in the initial list under the Manage Command. Click File, Save, or the Save icon to update the saved version of the map.

Create a map of home maintenance expenditures with the Thematic Mapping wizard

You also wish to examine the level of home-related expenditures in the Twin Cities area. To do so, you will create a thematic map displaying the average home-maintenance materials purchases per homeowner. You will use the normalization option of the Thematic Mapping wizard to do so.

1.   Open the Thematic Mapping wizard, select Block Groups as the layer to symbolize, select the Set the color of each area based on this attribute option, and Own Maint & Remodel Matls:Tot as the attribute to be mapped. You will find this attribute near the end of the list in the CY Consumer Spending category. (Note: As you become more familiar with Business Analyst data attributes, you can expedite this selection by clicking on the drop-down box for the attribute field, then entering text in the Filter box until the desired attribute is displayed. Click it to select it.) Click Next.

Review the list of consumer expenditure attributes. Several of them include purchases of goods and services that will be offered in Living in the Green Lane's store. The Thematic Mapping wizard does not provide the option to aggregate these attributes, you will do so manually later in this chapter.

2.   Select a color scheme from the available options. Select Quantile as the classification scheme. In the normalization field box, select CY Owner Occupied HU.

Consider these settings for a moment. You want this map to display expenditures per household rather than in total because you want to identify households with high home maintenance and remodeling purchases. However, this category reports expenditures for homeowners, not all households. Therefore the proper normalization attribute is CY Owner Occupied HU rather than Total Households.

3. Click Finish to apply these settings and create the map. It should resemble the one below, though your color scheme might vary.

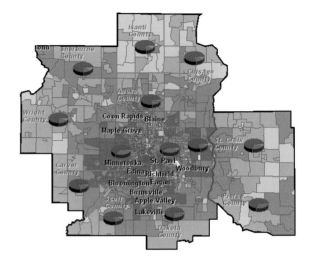

The map displays the desired data, but notice the legend in the Table of Contents. The labels are long and confusing. You will revise them by adjusting the layer's properties.

4. Right-click the Block Groups layer and select Properties (or double-click the layer) to open the Layer Properties window. Click the Symbology tab. Left click the Label button below and to the right of the Color Ramp box. (Do *not* click the Labels tab at the top of the window. If you do, click Symbology to return.) Select Format Labels. In the resulting dialog box, select Currency as the number format. Click OK. Click OK again to close the Layer Properties window and display the changes in the legend. To make this layer available for future use, you will save it as a layer file.

5. Right-click the Block Groups layer. Click Save as Layer File. Navigate to C:\My Output Data\Projects\LITGL Minneapolis St Paul\CustomData\ChapterFiles\Chapter2 and save the layer file as **BGHomeExpenditures.lyr**. You may now add this layer to this or any other map document by clicking the Add Data button, navigating to this folder, and double-clicking on this layer file.

6. Save the map file again to preserve your work.

You have now created maps displaying the patterns of home ownership, home value, and home-related expenditures per household in the Twin Cities area. Review these maps to identify concentrations of high home values and/or home-related expenditures. These represent areas of opportunity for LITGL. You must now create layers displaying the distribution of income and educational attainment to include these dimensions of the green-consumer profile.

Create a map of household income with layer properties options

In Business Analyst Desktop, thematic mapping may also be performed with Symbology options in the Layer Properties window as well as with the Thematic Mapping wizard. This approach provides greater control of thematic mapping options including classification and color schemes. You will use it to create thematic maps of household income at the block group level in the Twin Cities CBSA.

1. Turn off the County Areas layer. Right-click the Block Groups layer, click Properties (or double-click the layer) to open the Layer Properties window. Click the Symbology tab.

2. In the Value box, select CY Median HH Income. Click OK in response to the Maximum sample size message. To the right of this box, click the Classify button to open the Classification dialog box.

   The Classification dialog box provides significant information about the data values for this attribute. The top right of the box presents descriptive statistics for the attribute. Note that it includes 10,000 records, a sampling of the block groups in the book's dataset, which are, themselves, a small part of the more than 200,000 block groups in the United States. You may increase this sample size by clicking the Sampling key. The graph in the bottom left of the box displays the distribution of values for the attribute. The lines represent the class boundaries of the current classification scheme. The corresponding values are listed in the box to the right. Try moving one of the lines with your mouse or entering values directly into the list. The classification scheme changes to Manual and the class boundaries are adjusted accordingly. This allows you to customize class boundaries to specific values of importance to your project. By clicking the Mean and Standard Deviation options, you may elect to display these statistical measures as lines in the distribution chart.

   Note the Classification Method and Classes boxes at the top left of the dialog box. These allow you to set the classification method and number of classes for the map. When you change these settings, the revisions are immediately reflected in the chart and Break Values boxes. Experiment with different methods and numbers of classes to see the impact of these settings. You may even wish to click OK and Apply to see the impact of these changes on the map. You wish to display median household income in a quartiles method, with half the features above the median feature value and half below.

3. Select Quantiles in the Methods box and 4 in the Classes box. Click OK to return to the Symbology tab of the Layer Properties window.

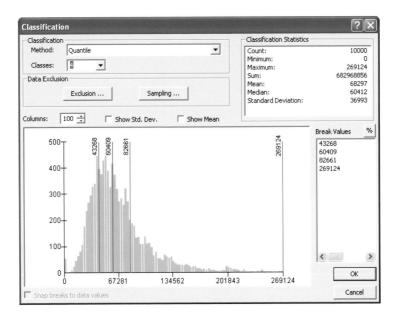

4.   In the Color Ramp box, select a Green to Blue color scheme. Note that attribute values are listed from highest to lowest. As you wish the higher values to be displayed in green, click the Range column header and click Reverse Sorting.

5.   Click the Labels column header to the immediate right of the Range column header and click Format Labels. In the resulting dialog box, click Currency, then OK. Click the Show class ranges using feature values box below the symbol column to use actual values in the data as class boundaries. The Symbology box should resemble this:

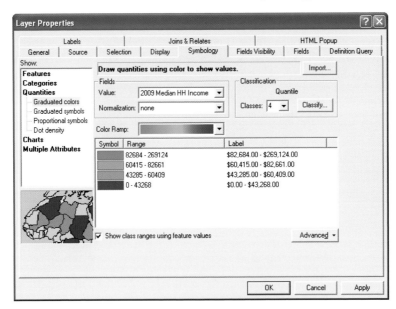

6. Click OK to close the Layer Properties box and apply the settings to the map. Your map should resemble the one below.

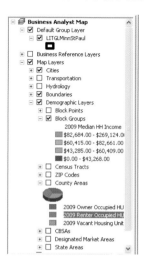

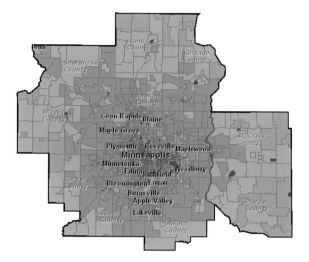

Review this map for a moment. The block groups in the map are not evenly distributed among the four classes even though you specified a quartile method, which should have produced that result. This is because the Thematic Mapping wizard (all block groups) and the Layer Properties classification scheme (a sample of block groups) use national collections of block groups in the United States (or, in this case, all the block groups in this book's dataset, not simply those in the study area) to determine class boundaries. Thus the classes displayed on the map represent the block group values related to national classes, not ones calculated for the Twin Cities area. As you wish to distinguish among block groups in this area, you wish to apply the quartile classification scheme solely to them.

To accomplish this task, you must use the data manipulation capabilities and the advanced thematic mapping tools in Business Analyst Desktop. These topics are the focus of chapter 3.

Chapter 3

Advanced thematic mapping and symbology; creating datasets and dynamic ring analysis

While the automated thematic mapping tools and layer property settings of Business Analyst offer significant capability, it is sometimes necessary to extend their power by customizing datasets and using additional symbology resources. In this chapter, you will learn to perform these tasks and use the datasets you create in conjunction with business data to understand Living in the Green Lane's business environment more fully.

Run Business Analyst Desktop and display map

1. Click Start, Programs, ArcGIS, Business Analyst, BusinessAnalyst.mxd to run ArcMap, load the Business Analyst Extension, and then load the default Business Analyst map.

2. Click OK when the Automatically update map projection dialog box appears, then close the Business Analyst Assistant window on the right of the screen.

3. Click File, Open. Navigate to C:\My Output Data\Projects\LITGL Minneapolis St Paul\ CustomData\ChapterFiles\Chapter3\LITGLBusinessPlan.mxd. Click the map file to open it.

Create a customized data source and use it to map household income

The map is the same one you designed at the conclusion of chapter 2 and resembles the one below. At the conclusion of that chapter, you noted the limitations of this map in representing the characteristics of the Minneapolis-St. Paul metropolitan area.

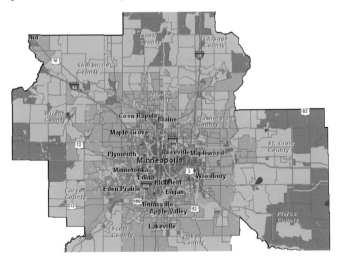

To view characteristics of block groups within the target area, you wish to define quartile groups based only on that area rather than on national distributions. To achieve this goal, you will create a new shapefile containing only the block groups in this area and only the attributes you wish to include in this analysis. This solution will also allow you to aggregate the home-related expenditure categories you identified above. You will also apply this approach to educational attainment attributes to customize a new attribute to display on a thematic map.

Business Analyst Desktop facilitates the creation of customized datasets through a combination of selection operations and layer properties settings. You will use both to create a block groups dataset with the attributes you need to create the customized thematic maps you seek.

1. Open the Layer Properties window for the Block Groups layer. Click the Fields Visibility tab. This box designates the attributes that will be visible for each feature. By default, all the attributes in the dataset are displayed. These are listed in the Visible Fields box on the right. Click the double left arrow button to move them all to the Invisible Fields box on the left.

2. Press the Ctrl key and select attributes from the list on the left by clicking them. The Ctrl-click combination allows you to select multiple attributes individually. The Shift-click option allows you to select a range of adjacent attributes by clicking the first and last attributes in the list. Remember that the CY designation in the attributes below refers to the current-year data in your Business Analyst installation, while FY refers to future-year projections. When you have selected the attribute or attributes you want, click the single right arrow to add them to the Visible Fields list. (Careful—if you click the double right arrow, you will move all attributes to the Visible Fields list. If that occurs, reverse this operation and repeat the selection process.)

3. Click the Manage button below the Used Saved Variable List option to open the Manage Lists box. Select LITGLIncEdOwnExp from this list, then click Apply. Confirm that the attributes in the list below are now included in the Visible Fields list, then click OK to close the box. If you wish, open the attribute table to verify that the following attributes are the only ones displayed:

OBJECT ID
SHAPE
ID
CY Total Households
CY Average Household Size
CY Total Housing Units
CY Owner Occupied HU
CY Renter Occupied HU
CY Median Age
CY Pop Age 25+ Educ Attain Base
CY Pop Age 25+ by Educ: Assoc Deg
CY Pop Age 25+ by Educ: Bach Deg
CY Pop Age 25+ by Educ: Grad Deg
CY Median HH Income
CY Median Value: Owner HU
FY Total Housing Units
FY Owner Occupied HU
FY Renter Occupied HU
Dominant Tapestry Code

Home Imp Services-Own & Rent: Tot
Home Imp Materials-Own & Rent: Tot
Major Appliances: Tot
Small Appliances: Tot
Lawn and Garden Tot

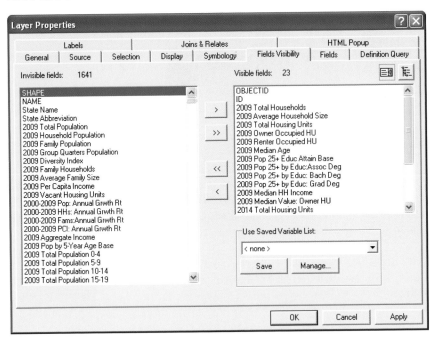

4. Click the OK button. You have now designated the attributes for the new dataset. You will use a Select by Location operation to designate the features.

5. In the menu bar, click Selection, then Select by Location to open the Select by Location dialog box. Select the necessary options to specify that you wish to select features from Block Groups that have their centroid in features in the LITGLMinnStPaul layer. Click OK to close the box and select all the block groups in the LITGLMinnStPaul layer.

The map now resembles this, with all the block groups displayed as being selected.

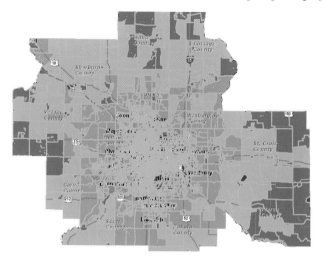

6. Right-click the Block Groups layer, then click Open Attribute Table. When the table opens, notice that it displays only the attributes you designated. Further, the text at the bottom of the table lists the number of selected features. Click the selected button to display only those features. This is the combination of features and attributes you wish to include in the new shapefile. Use the button at the top right of the box to close the attribute table.

7. Right-click Block Groups layer, click Data, then Export Data to open the Export Data dialog box. In the Export box, select Selected features. Select the same coordinate system as the layer's source data option. In the output shapefile or feature class box, click the Browse (yellow folder) icon, navigate to the C:\My Output Data\Projects\ LITGL Minneapolis St Paul\Custom Data\ChapterFiles\Chapter3\ folder and enter **BGIncEdOwnExp.shp** as the filename and select Shapefile in the Save as Type box. Click Save. The Export Data dialog box should resemble this:

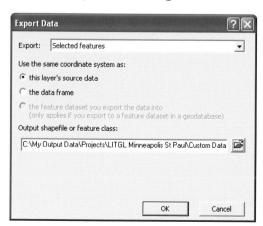

8. Click OK to export the selected features and attributes to a new shapefile in the designated folder. Respond Yes when given the opportunity to add it to the map as a layer.

   The new layer is added at the top of the data frame. You will move it to a better position in the Table of Contents, and use it to create a new thematic map displaying the distribution of median household income by quartiles in the Twin Cities area.

9. Confirm that the Display tab is selected at the lower left of the Table of Contents. Click and drag the BGIncEdOwnExp layer down to the first position in the Demographic Layers group layer in the Table of Contents.

10. In the menu bar, click Selection, then Clear Selected Features to remove the block group selection.

11. Open the Layer Properties window for the BGIncEdOwnExp layer. Click the Symbology tab. Reproduce the quartile classification settings (Quantile classification with four classes) for the household income thematic map you created below. The Symbology window should look like this:

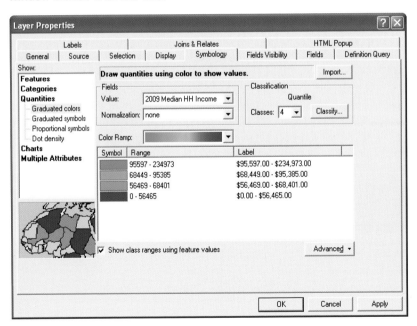

12. Click the General tab and enter **Median HH Income by Block Group** as the map title. Click OK to close the Layer Properties window and apply the settings to the map, which should resemble this:

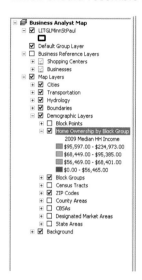
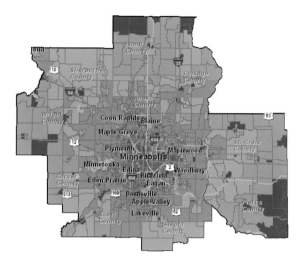

Compare this map to the household income map you created earlier. Block groups are more evenly assigned to classes, reflecting a classification scheme calculated solely from Twin Cities area data. This map is more useful to you in identifying the concentrations of upper income households associated with the green-customer profile.

13. Save this layer as a layer file, **BGHHIncome.lyr**, in C:\My Output Data\Projects\LITGL Minneapolis St Paul\Custom Data\ChapterFiles\Chapter3. Use the Save As command to save your map file as **LITGLBusinessPlan2.mxd** in the same chapter 3 folder to preserve your work.

Create and calculate new fields for thematic mapping

To this point, you have symbolized important characteristics of the green-consumer profile with pie charts, data attributes, and normalized data attributes. However, the characteristic, educational attainment, is not captured in a single attribute, but dispersed over a range of related attributes. This is also true for home-related expenditures, which you also wish to display in a single thematic map. Thus, you must perform calculations involving multiple attributes to produce the required values. To do so, you will add fields to your new shapefile and calculate the attributes required for these maps.

1. Right-click the Household Income by Block Group layer, then click Open Attribute Table to open the layer's data table.

2. Click Options, then Add Field to open the Add Field dialog box, enter **TotCollDeg** (Total College Degrees) in the Name field and select Double as the data type. When your screen resembles the one below, click OK to create the field and close the dialog box.

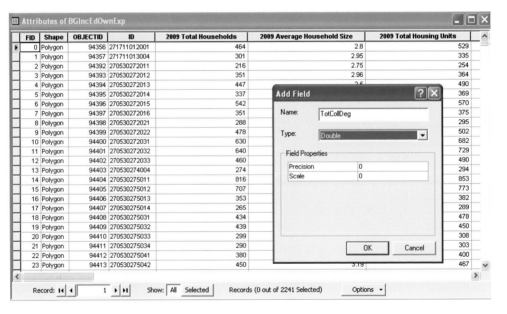

3. Repeat this procedure to create a second new field **TotHmReExp** (Total Home Related Expenditures) with a Double data type.

4. Scroll to the far right of the attribute table where you will find the new fields with values of 0 for all features. You will use these fields to calculate the correct values for each block group in the dataset.

5. Right-click the TotCollDeg label at the top of that field's column, then click Field Calculator. Click Yes to the message about calculating outside an edit session. The Field Calculator box opens.

The Field Calculator box allows you to create an expression that will calculate the values in the new field for each block group. You wish to calculate the total number of people over 25 with a college degree in each block group. To do so, you must sum the number of people with associate, bachelor, and graduate/professional degrees. You will use the Field Calculator box to build this expression.

6. Double-click the EDASSC_CY field to add it to the expression. Click the + button (or press the + key on your keyboard) to add the operator. Double-click EDBACH_CY, click +, click EDGRAD_CY. These attributes contain the number of people 25 or older with associate, bachelor and graduate/professional degrees. When the expression resembles the one below, click OK to calculate the values and add them to the attribute table.

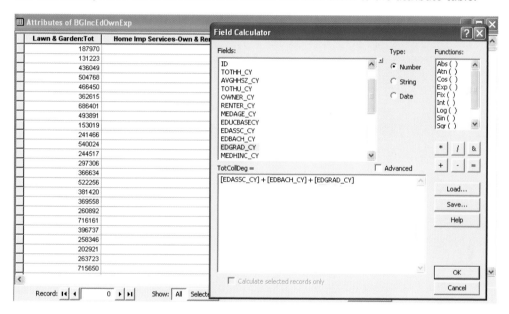

7. Repeat this process on the TotHmReExp field, using the Field Calculator to build the following expression for this attribute.

   [X4060_X] + [X4080_X] + [X4009_X] + [X3047_X] + [X3048_X] (These are the five expenditure categories you designated in the shapefile.)

8. Click OK to calculate the field, then close the attribute table.

   The calculated fields are ready for use in thematic mapping and other operations. However, the field names are a bit cryptic and may not convey their contents to other users. You will add aliases for both fields to clarify their contents.

9. Open the Layer Properties box for the Household Income by Block Group layer, then click the Fields tab. Scroll to the bottom of the attribute list to find the two new fields. Note that their Name and Alias are identical. Click the Alias field for TotCollDeg attribute and enter **2009 Total College Degrees** as the Alias. Repeat this procedure to enter **2009 Total Home Related Expenditures** as the Alias for the TotHmReExp attribute. When the Fields window resembles the one below, click OK.

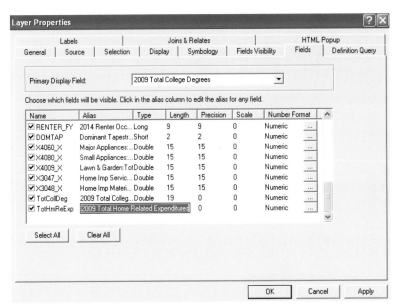

The aliases are added to the new fields and will now be displayed at the top of the fields in the attribute table and in the field list in the Symbology window. If you wish to see the Alias fields, open the layer's attribute table and scroll to the far right. The new fields may now be displayed in thematic maps.

Create maps of educational attainment and home-related expenditures

The new fields in the shapefile allow you to produce thematic maps depicting Educational Attainment and Home Related Expenditures both in total and normalized with the proper field for each attribute.

1. Open the Layer Properties window for the Household Income by Block Group layer. Click the General tab and enter **Educational Attainment by Block Group** in the Layer Name box. Click the Symbology tab, then press the Quantities tab at the left of the Symbology screen. Select 2009 Total College Degrees as the Value. Click Apply to apply the settings and display the thematic map.

The map refreshes behind the Layer Properties window to display the new settings. Each block group is now displayed in a color which represents the number of people over 25 in the block group who hold a college degree. Though this is useful, you are more interested in comparing the percentages of people with college degrees by block group. To do so, you will normalize the thematic map as you did above, using the attribute 2009 Pop25+ Educ Attain Base. This attribute reports the number of people in the block group who are 25 and older and is the value to which all the individual educational attributes sum.

2. Select 2009 Pop25+ Educ Attain Base as the Normalization field. Note that the ranges and labels now change to decimals. To display these values as percentages, click the Label button just above the decimal values to open the Number Format window. (Note: Do *not* click the Labels tab in the Layer Properties window. If you do so in error, click the Symbology tab to return to the correct window.)

3. Select the Percentage option in the Category box; select The number represents a fraction option. Adjust it to show a percentage option, then click Numeric Options. In the resulting box, designate the Number of decimal places as 1, then click OK twice.

4. Select the Show class ranges using feature values near the bottom of the Symbology window. Confirm that the Classification scheme is set to Quantile with 4 classes. When the window resembles the one below, click OK to close the window, apply the new settings and refresh the map.

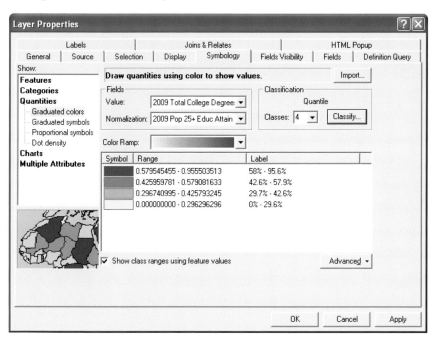

5. Right-click the Educational Attainment by Block Group layer, click Save as Layer File and save the file as **BGEdAttain.lyr** in C:\My Output Data\Projects\LITGL Minneapolis St Paul\CustomData\ChapterFiles\Chapter3\.

You have now designed maps for each of the major population characteristics of the green-customer profile. You will complete the thematic mapping task with a map that displays average household expenditures on home-related products and services by block group.

6. Repeat step 1 above by entering **Home Related Expenditures per HH by Block Group** in the Layer Name box of the General tab and duplicating the settings in the graphic below in the Symbology tab. Be sure to select a new color scheme. Click OK to apply the settings and display the Home Related Expenditures per HH by Block Group thematic map.

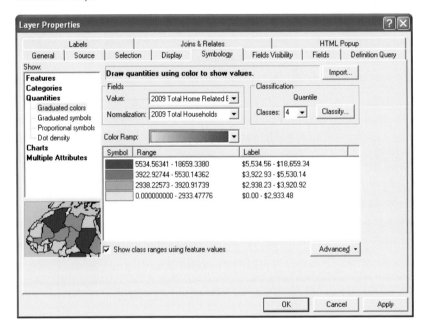

7. Right-click the layer, click Save as Layer File and save the file as **BGHomeExpHH.lyr** in C:\My Output Data\Projects\LITGL Minneapolis St Paul\CustomData\ChapterFiles\Chapter3\.

8. To add any of the layer files you have created to the Table of Contents, click Add Data, navigate to C:\My Output Data\Projects\LITGL Minneapolis St Paul\CustomData\ChapterFiles\Chapter3, select the layers you wish to add, click Add. Turn the layers on and off in the Table of Contents to view the distribution of each attribute.

You now have a collection of layer files capturing different population characteristics and home-related spending patterns in the Twin Cities area. You will now create maps of retail shopping centers and competing home centers to display in conjunction with your thematic maps.

Create maps of shopping centers and competing home centers using point features

Shopping centers and businesses are displayed as point features in Business Analyst desktop. Successful retail shopping centers attract significant customer traffic to their locations, increasing their attractiveness as potential sites for LITGL's first store. Existing home centers, on the other hand, are direct competitive threats for sales. To advance your understanding of the Twin Cities business environment, you will display the location and relevant information about these point features on your map.

1. Adjust the layers in the Demographic Layers group layer to display your original thematic map of household income. Turn on the Business Reference group layer and the shopping centers layer. As this layer is scale dependent, you must zoom to the central region of the study area to display the shopping centers layer. Continue zooming until this layer becomes visible. If you wish, toggle other Map Layers to make shopping center locations more visible. Your map should resemble this one, though it may be zoomed to a different region of the market area.

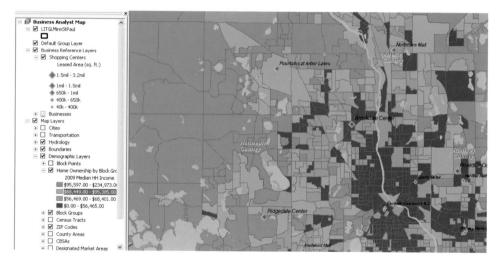

2. Expand the Business Analyst Toolbar, click Data and Maps, then click Add Business Listings to open the Add Business Listings window. Click the Select button to the right of the Location box to open the Select Location dialog box. Select the Search inside the features of the polygonal layer specified option and select LITGLMinnStPaul as the layer to be searched. Click OK.

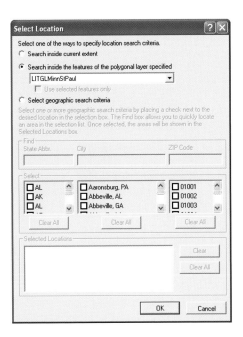

3. Click the Select button to the right of the Type of Business (NAICS) to open the Business Type Dialog box. In the Key Words box enter **home center**. Notice that as you type, the selection available to you is adjusted to match your entry. Select the first HOME CENTERS option. The dialog box should resemble the one below. Click OK.

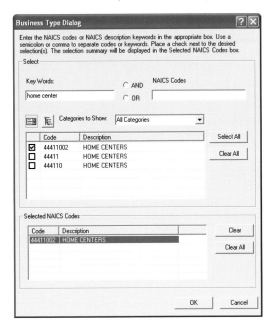

The Add Listings window should appear like the one below. Your query will extract from the Business Analyst desktop database those businesses designated as Home Centers in the Twin Cities CBSA.

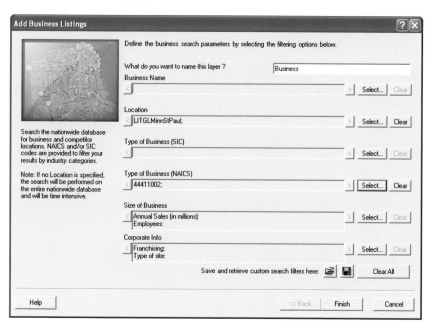

4. Click Finish. The listings are extracted and added as a Businesses layer to the group layer at the top of the Table of Contents.

   Janice and Steven believe that annual sales are the best measure of the competitive strength of a competing home center. You will adjust the symbology of the Businesses layer to reflect that information.

5. Open the Layer Properties box for the Businesses layer. Click the Symbology tab. In the Show box, select Quantities, Graduated Symbols. Select SALES_VOL in the Value box. Click the Classify button and choose three Classes and a Quantile classification Method. Click OK.

6. Click the Template button to open the Symbol Selector dialog box. Click the More Symbols button at the bottom right of the box. This opens a list of collections of available symbols. Click the Business option to make these symbols available. Scroll approximately three-quarters of the way down the set of available symbols until you find the House2 symbol. Select it and set its color to Red. The Symbol Selector dialog box should resemble this. Click OK to return to the Symbology tab.

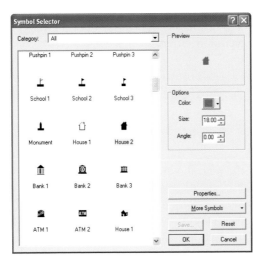

7. In the Symbol Size from and to boxes, set **12** and **18** as the low and high limits. Format data labels to currency and sort the symbols from large to small and the labels from high to low, with the largest symbols and highest values at the top of the legend. Select the Show class ranges using feature values option. Click the General tab and enter **Home Centers by Sales in 000's** as the map title. Click OK to apply the settings and display the map. Zoom into the center of the map until the Shopping Centers layer displays as well. Your map should resemble this:

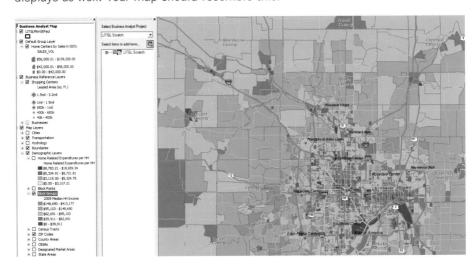

This map displays shopping centers by size of leased area and home centers by volume of annual sales. Integrating these layers with your thematic maps allows you to explore their locations and concentrations of attractive population characteristics simultaneously.

Integrate population and business maps to study competitive environment

The final step in this chapter is to integrate the map layers you have created, enabling you to study the relationship between the population and business layers. You will use the population layers as base maps and display the business layers above them.

1. In the menu bar, click File then Add Data (or click the Add Data button) ✚.
    If necessary, Navigate to the project folder containing the layer files you have saved.
    Press the Ctrl key and click on BGHomeExpHH.lyr, BGEdAttain.lyr, and BGHHIncome.
    lyr to select them. Click OK to add them to the Table of Contents.

    The new layers display correctly but obscure the business layers beneath. To correct this move them below the business layers in the Table of Contents.

2. Press the Ctrl key. In the Table of Contents click on the new layers individually to select them. Drag this collection of layers to just above the Block Groups layer in the Demographic Layers group layer. Contract the layers and group layers you are not using to see more active layers.

3. Save your map file to preserve your work.

You map should resemble the one below. It displays the Shopping Centers and Home Centers layers above a thematic map of home-related expenditures.

Turn the thematic layers on and off to adjust their visibility. For each layer, explore the distribution of shopping centers and home centers relative to concentrations of attractive population characteristics. Areas of underserved households that match the green-consumer profile represent potentially favorable locations for Living in the Green Lane's first store. You will use the Dynamic Ring Analysis Tool to identify these locations more accurately.

Identify attractive locations with the Dynamic Ring Analysis Tool

1. In the menu bar, click Window, then Table of Contents to turn it off. Notice these two buttons on the Business Analyst desktop toolbar. The red bull's-eye on the left is the Site Prospecting tool and the red button on the right is the Dynamic Ring Analysis Tool.

    These tools enable you to explore areas that appear attractive on a map. You will use the Dynamic Ring Analysis Tool, the red one on the right, first.

2. Click the Dynamic Ring Analysis Tool . This loads a graph displaying selected attributes on the left of the map. Click the Change Parameters button at the bottom of the graph to load the Dynamic Ring Analysis wizard. Select Standard Business Analyst Data (Generalized) in the data layer box. Remove the default fields from the Select fields box. Using the dropdown box in the Select fields box, select the following attributes and add them to the Select fields box:

    CY Total Households
    CY Owner Occupied HU
    CY Pop Age 25+ by Educ: Assoc Deg
    CY Pop Age 25+ by Educ: Bach Deg
    CY Pop Age 25+ by Educ: Grad Deg

    The Dynamic Ring Analysis box should look like this.

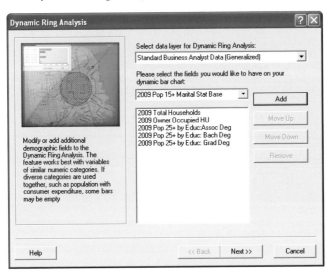

3. Click Next. In the resulting box, Select the option to Set the threshold field and value, select Shelter:Tot as the threshold field and enter **150000000** ($150 million) as the threshold attribute. This is the broadest measure of consumer spending on housing and includes a number of other expenditure categories relevant to Living in the Green Lane. Set the Radius as 2 and the Distance unit as Miles. With these settings the Dynamic Ring tool will distinguish between those 2-mile ring areas in which total expenses on shelter are greater than $150 million and those which fall below this level. This box should resemble the one below. When it does, click Finish to apply the settings and close the Dynamic Ring Analysis dialog box.

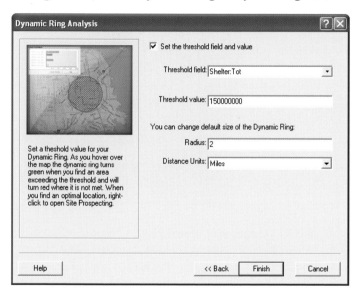

4. With the Dynamic Ring icon, move around the map, zooming in if necessary to view its features more clearly. Locate a site that is conveniently located relative to the freeway system, near block groups with high levels of home-related expenditures and relatively close to shopping centers but removed from competitors. This is a relatively complex search (think of writing the SQL statement) but the visual clarity of the map makes it fairly simple. When you have found a suitable location, click on it. Depending on where you clicked, your map will resemble this one in structure, but not in location.

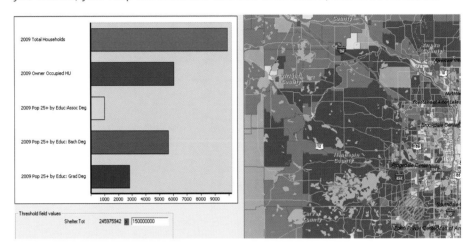

The map displays a hashed circle in a 2-mile ring around your chosen site. A red ring indicates that this area falls below the $150 million threshold level of Total Shelter Expenditures, while a green ring indicates an area above the threshold. The bar chart to the left displays the number of households and owner-occupied housing units in the area, and the number of adults in each of the three educational attainment attributes indicating a college degree.

Consider for a moment, the rich data context in which you are working. All the major characteristics of the green-consumer profile are represented here, as are consumer expenditures in the categories most relevant to LITGL. In addition, shopping centers, competing home centers, and the freeway system in the area are superimposed over the thematic map. In short, the various components of the Business Analyst system combine to create a very comprehensive, yet easily comprehensible view of Living in the Green Lane's competitive environment.

Continue to explore this environment, sampling several locations for a favorable combination of factors. When you have found a site you like, proceed to the next step.

5. Right-click the site you selected, click Prospect selected point to open the Site Prospecting Wizard. (This could also be done with the Site Prospecting Tool 🔍.) Set the options in the wizard to create a single 2-mile ring market area around that site. Leave the Remove Overlap and Donut options unselected, then click Next.

2

3

6.   Name the new Trade Area **2Mile Ring**. Select Create reports, click Next. Select the For Individual Features option, click Next. Select the Comprehensive Report, click Next. Select the option to View reports on screen. (If Business Analyst does not display the available report templates, click the Options button and select Standard BA Data as the layer you want to summarize.) Click Finish to create the market area, place it on the map, generate the selected report, and display it on screen. When you zoom out your map will resemble this one though the location will be different.

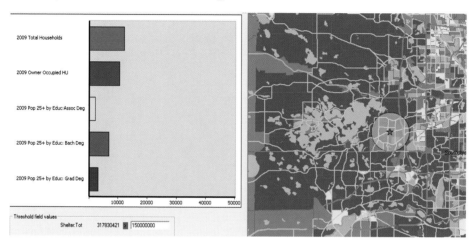

Review the Comprehensive Report relative to the desired population characteristics and purchasing patterns. It includes overall spending on the Shelter category as well as several other categories of home related expenditures relevant to Living in the Green Lane. Note how this process resembles the market area definition and report generation functions of Business Analyst Online. Note as well the similarity between reports and contents in the two systems.

As you explore areas for potential stores, you wish to supplement these tools with satellite imagery to further enhance your understanding of each location.

Add satellite imagery with an ArcGIS Online layer

ArcGIS Online provides a wide range of ready-made maps which you can add to your work. This function is fully integrated into the ArcGIS system and is accessible, from the Maps button on the main menu **Maps ▼**. You will use this tool to add satellite imagery for the site described in the Comprehensive Report.

1.   Click the Maps button, and review the options, which will resemble those offered below.

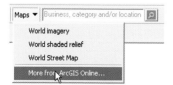

2.   Click the More from ArcGIS Online option, then click OK to the explanatory text to open your browser and navigate to the ArcGIS Desktop Resource Centers Web site, which resembles this:

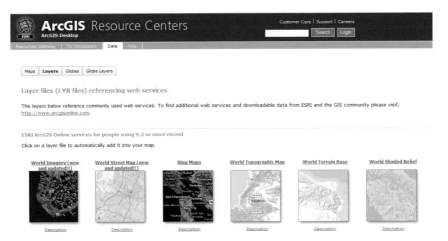

The Layers tab is open displaying a variety of map layers directly available for your work. The site also provides directions for integrating these resources into your maps. You may also explore the other tabs for additional features of the Resource Center. When you are finished, close this site and return to Business Analyst, where you will load the World Imagery layer directly from the Maps command.

3.   In Business Analyst, click the Maps button, then select World Imagery. Business Analyst loads this map layer and adds it to the bottom of the Table of Contents. Turn the upper layers off as needed to view the World Imagery layer. Zoom to the site you prospected in the Comprehensive Report. Your map should display the trade area over satellite imagery of the area you selected.

Zoom further to different parts of the 2-mile ring around your selected site. (You may wish to turn off the ring layer when doing so.) Note that the resolution of the layer adjusts to the map's scale, revealing more detail as you zoom in. This is done automatically as more detailed imagery is accessed on the ArcGIS Online server in response to your commands. If you have difficulty with this function, check your Internet connection, as it is the channel through which this data is routed. This imagery allows you to explore the area of interest in greater detail and to view characteristics that are not captured in the maps you designed.

All these processes may be used to identify attractive sites. However, they do not reveal whether actual retail sites are available within those locations. The next part of the book will add that data, integrate it with the work you have done and use the results to select Living in the Green Lane's first store location from among those options.

ROI considerations

The tasks in these chapters are essential components in the process of selecting the site for Living in the Green Lane's first green-home center. For this reason, the ROI impact of these analyses is included in the discussion at the conclusion of Part III.

Summary of learning

In chapters 2 and 3, you have developed your thematic mapping skills and used them to develop a comprehensive understanding of the population and business dimensions of Living in the Green Lane's competitive environment. In the next part of the book, you will focus your attention on selecting the best retail site available within that environment for the company's first store.

Before moving on to this task, however, reflect for a moment on what you have learned in these chapters.

You have expanded your GIS knowledge by learning:

1. The value and basic design features of thematic maps, including point symbolization
2. The types of data available in Business Analyst Desktop and how to select appropriate measures for specific business analysis
3. The value of normalization, when it is appropriate and some common normalization attributes
4. The classification schemes available in Business Analyst and how they affect the display of attribute data
5. The value of manipulating standard demographic data to create attributes relevant to specific business problems
6. The value of integrating multiple data sources into a comprehensive, comprehensible map document
7. The value of using imagery resources to understand areas of interest more fully

You have enhanced your GIS skills by using Business Analyst to:

1. Create a study area for a business environmental mapping project
2. Design graph maps and thematic maps with the Thematic Mapping Wizard and Layer Properties options
3. Use Layer Properties options to map point feature attributes
4. Extract custom datasets for specific business problems
5. Create customized shapefiles, calculate new attributes from existing ones, and use them to create thematic maps
6. Integrate layers displaying population characteristics with business information layers to create comprehensive maps of an enterprise's business environment
7. Use the Dynamic Ring and Site Prospecting tools to discover favorable potential retail locations
8. Access imagery layers from ArcGIS Online and integrate them into map documents

Notes
1. The following two sections rely heavily on Cynthia Brewer's *Designing Better Maps: A Guide for GIS Users* ESRI Press, Redlands, Calif. 2005.

Part III

Trade area analysis and site selection without customer data

Relevance	Entrepreneurial new businesses must convince investors that they have selected the most favorable site for their new ventures.
Business scenario	LITGL must demonstrate to potential investors that it has selected the best site for its first store given the population characteristics and competitive environment in which it will operate.
Analysis required	LITGL must identify properties available for its first store, geocode their locations, and integrate them into a GIS project. LITGL also must evaluate the locations relative to concentrations of attractive potential customers, shopping centers, and competing home centers, as well as create trade areas around them, create general and detailed reports for each trade area, then use all this information to select a site for the first store and justify this selection in its business plan.
Role of business GIS in analysis	Integrate external data on available locations into GIS project. Create trade areas and generate reports for them. Use reports to select most attractive location. Create map documents to communicate and support recommendation.
Integrated business GIS tool	ESRI Business Analyst Desktop 9.3.1 and ESRI Segmentation Module extension.
ROI considerations: Cost of business GIS	Business Analyst Desktop and Segmentation Module purchase, salary of GIS analyst.
ROI considerations: Benefits of business GIS	Increased sales of recommended location versus alternate available properties. Potentially lower loan costs if lender understands reduced risk from extensive site selection analysis.
Environmental impact of business decision	Lower driving time and impact for most responsive customers. Select site with appropriate characteristics to demonstrate and sell environmentally clean technologies.

Table III.1 Executive summary

The Living in the Green Lane scenario

Janice Brown and Steven Bent's business plan is nearing completion. The business model is clearly defined, target customers identified, and their distinctive characteristics profiled. In the previous two chapters you used ESRI Business Analyst Online and ESRI Business

Analyst Desktop to reveal concentrations of target customers in the Twin Cities area. You have also examined Living in the Green Lane's competitive environment by exploring the distribution of shopping centers and competitive home centers relative to the Minneapolis-St. Paul transportation infrastructure. Using both Business Analyst Online and Business Analyst Desktop, you have identified attractive concentrations of underserved target customers.

The final element of LITGL's business plan is the selection of a specific location for the company's first store. That is your task in chapters 4 and 5. To complete it, you must evaluate the attractiveness of a set of available sites provided by a commercial real estate agency. You must match these sites to the criteria established by Janice and Steven and seek out the site with the best location for serving targeted customers and competing successfully. You will use the market area creation and reporting functions of Business Analyst Desktop to accomplish this task in six steps.

First, you will add the list of available properties to your Business Analyst project and geocode the locations to display them on a map. You will then examine the attributes of each location and identify the ones that meet Janice's and Steven's selection criteria. Specifically, they are seeking an existing freestanding retail facility of 40,000 to 60,000 square feet. This is a relatively compact size for a home center, but Janice and Steven believe that a smaller store size is consistent with their environmental vision and would serve a smaller, but more attractive market area.

Janice and Steven also require ample display and warehousing capabilities and substantial space for parking and outdoor demonstrations. Specifically, they wish to have four or five parking spaces per 1,000 square feet of retail floor space. They plan to convert the facility to a green building with green parking. This would illustrate the benefits of their business concept and create a comparatively modest footprint for a retail site, improving opportunities for replicating the facility in other neighborhoods. These criteria will guide your selection.

Second, you will use the customer prospecting and market area functions of Business Analyst Desktop to compare the locations of available properties with concentrations of attractive customers and the competitive environment of the area.

Third, you will create equal probability trade areas around competing home centers and create Market Locator Reports for each available potential site to explore the competitive environment of each.

Fourth, you will create drive-time polygon trade areas around each available property and create a set of reports for each using the reporting function of Business Analyst Desktop. You will select reports relevant to the enterprise's target customer profile, as well as consumer expenditure and Tapestry Segmentation lifestyle segmentation patterns in the market area of each qualified location.

Fifth, you will use the maps and reports you have created to select the site for the first store offering the most attractive combination of proximity to targeted customers within a favorable competitive environment.

Finally, you will use the layout capabilities of Business Analyst Desktop to design map documents for the business plan report that supports your conclusions and recommendation.

Integrated business GIS tools in trade area analysis and site selection

Selection of the first site for a new retailing business is a difficult process. There is no successful store to serve as a model and no data on existing customers from which to develop a profile. Competitors are well-established and the new company faces the challenge of disrupting habitual buying patterns. On the other hand, traffic flows and concentrations of retail centers also are well-established, presenting opportunities for reaching prospective customers quickly.

Companies face this challenge by defining their target customers precisely, identifying their most direct competitors clearly, and understanding the competitive environment as fully as possible. For Living in the Green Lane, the green-customer profile provides a clear target market profile. Concentrations of households with matching characteristics are clear market opportunities.

The company also must decide which population characteristics to consider: residential or daytime. Many retail stores, especially in urban areas, serve customers on the way to work or back or at other times during the work day. As many of these customers commute and live elsewhere, data on the residential population near the store's location is significantly less relevant. Janice and Steven believe that this is not generally the case for home center shoppers and will be even rarer for Living in the Green Lane's customers. Thus, residential population statistics are more relevant.

Identifying competitors can also be a challenge, given the breadth of merchandising strategies in modern retailing. For example, consumers can buy building supplies from a variety of sources including building supply companies, hardware stores, mass merchandisers, home centers, and online retailers. Janice and Steven believe that home centers, which offer a combination of extensive inventory, knowledgeable sales staff, and installation support, will be the most direct competitors to Living in the Green Lane's business model.

Transportation infrastructure and retail centers are also relevant considerations. As you observed in the previous chapter, shopping centers in the Twin Cities area are located largely in the vicinity of major highways. In addition, most existing home centers are located in relatively close proximity to shopping centers. Their objective is to attract customers drawn to these retail centers to their freestanding stores nearby. You will explore similar possibilities for Living in the Green Lane's first store.

Integrated business GIS functions support site selection decisions by integrating and customizing population, retail, competitive, and infrastructure data. Specifically they allow users to create trade areas for their own stores, competing stores, and available sites. The characteristics of each trade area are then captured in a series of reports which, collectively, provide a comprehensive view of the relative advantages and disadvantages of potential locations. The breadth of potential trade area definitions and preformatted reports provide a solid foundation for the first store site selection process. This functionality provides users with very powerful tools for business decision making, tools that previously were the exclusive domain of large, resource-rich organizations with specialized GIS analysts.

Trade area options

You worked with three distance-related trade area approaches—rings, donuts, and drive-time polygons—in the previous chapter. All of these methods are available in Business Analyst Desktop as well as Business Analyst Online. In these methods, distance is the primary defining factor for trade areas—linear distance in the first two approaches, and distance covered by drive times in the third. These approaches are useful for trade areas focused on residential population and influenced by transportation infrastructure and retail centers, as is the case here.

Data-driven polygons define trade areas based on the values of selected attributes. The grid approach is based on a set of important defining attributes, one of which is designated as the attribute to be symbolized. The system then creates a grid of equally sized cells over the relevant area, the color of each cell reflecting its value for the symbolized attribute. The data table for the grid contains aggregated values for the other selected attributes. This approach allows the user to identify, for example, square-mile areas within a region that have high home values and explore their relevant characteristics.

Though not technically a trade area method, you will use Business Analyst Desktop's Customer Prospecting tool in conjunction with the grid approach to identify concentrations of LITGL's most attractive prospective customers.

Threshold rings are a second data-driven method. This approach creates ring trade areas around locations based on user-defined values, or thresholds, for selected variables. In this approach, trade areas around a store site might be drawn to include, for example, 50,000 households within an inner ring and 75,000 households within an outer ring. You will use total shelter expenditures to draw threshold rings that display the concentration of these purchases around available sites in the Twin Cities area. This approach is useful when a specific level of population or expenditures is necessary to support each store in a market area.

Competition-driven trade areas are defined by characteristics of existing competitive stores. In the Equal Competition approach, that characteristic is linear distance. This approach draws trade areas in the form of Thiessen polygons around existing stores. Thiessen polygons are bounded by lines, each of which is composed of points equidistant between the nearest two points in a set of points. Thus each polygon defines the area that is closer to its central point than to any other point in the set. When applied to trade areas, this means that the boundary line between any trade area polygons is composed of points equidistant from the two nearest stores. Thus, if distance is the determining factor in choice of a store, the consumers within each store's trade area polygon would shop at that store, while those on the boundary line would be just as likely to shop at one of the two nearest stores as the other. Since distance is rarely the only determining factor and linear distance is unlikely to reflect the route distance to each store, this approach is illustrative but not definitive.

The Huff Equal Probability Trade Area approach overcomes some of the deficiencies of the Equal Competition approach by including additional factors in the creation of trade areas. Named for its creator, Professor David Huff, a marketing strategy researcher at the University of Texas McCombs School of Business, this approach acknowledges the influence of distance, but also accounts for other factors which affect consumer store choice. Based on consumer surveys and/or strategic considerations, users can define and weigh additional factors included with linear distance in the determination of trade areas. Such factors might include size of retail space in square feet, number of employees, number of product lines, and/or store sales. In this approach, the size of each store's trade area is related to these factors as well as to its linear distance from competing stores. This approach is appropriate for adjusting trade areas to reflect attractive competitive factors in retailing. You will use it to define trade areas for existing home centers that are Living in the Green Lane's major competitors.

However trade areas are defined, Business Analyst Desktop provides a full range of report templates to explore their characteristics. This information is vital in determining the most appropriate site for a new store.

Types of reports

The robust reporting function of Business Analyst Desktop provides a powerful tool for understanding the characteristics of trade areas, however they are defined. Users can select reports from an extensive list of existing templates, edit those templates to create customized reports, or build desired reports from the ground up. The report templates provided in the package are quite extensive. They include three categories of reports, each of which serves a different objective.

Summary reports

These are short reports, each of which provides a snapshot of trade areas relative to a specific characteristic or a focused set of general characteristics. Data is reported for the current year and five-year projections. Most of these reports include maps of the relevant trade areas as well. This category includes the General Report, Age Report, Race Report, Population Report, Multi-Area Report, and Household Report.

Comprehensive reports

These are longer reports, each of which contains a range of data on population characteristics, expenditure patterns, and/or Tapestry Segmentation lifestyle segment composition. Age, income, net worth, and home value distributions are provided in many of these reports. The Executive Summary provides a valuable collection of the most commonly used market measures. The Comprehensive Report includes all these categories as well as distribution tables for age, income, net worth, and home value. The Comprehensive Trend Report provides similar information, but expands many of the distribution tables with estimates for five-year projected values as well. Both include Tapestry Segmentation lifestyle data. The Market Profile Report provides basic statistics with current and five-year projects as well as a few basic distribution tables. It also provides a variety of data from the most recent census on educational attainment, housing, driving, employment, home ownership, and household composition patterns, among others. The Retail Expenditure Report includes data on several different categories of consumer expenditures. In each category, total expenditures and average household expenditures are reported along with an index that relates average household category spending in the trade area to the national average, with values above 100 indicating higher-than-average spending and values below 100 lower than average. The purpose of these reports is to provide a comprehensive picture of trade areas to inform site selection decisions. Indeed, you will use these reports for that purpose in this chapter.

Segmentation reports

Age, sex, and income are commonly used demographics for market segmentation purposes. This report category includes extensive distribution tables for various combinations of these measures. The Demographic and Income Profile Report includes distribution tables for age, race, and income from the most recent census, current year estimates, and five-year projections. It also presents this information graphically for the current year. The Age by Sex Profile Report, not surprisingly, contains distribution tables reporting population distribution by sex within several age range classifications.

Business Analyst Desktop's trade area definition and site reporting tools will provide you with the information base to support your site selection decision. In chapter 4, you will geocode the available properties and evaluate their locations relative to their business environment. In chapter 5, you will define specific trade areas, generate reports for them, select the best site, and design map documents to support your decision.

Chapter 4

Geocoding and evaluating alternative potential sites

The process of site selection begins with geocoding the addresses of available qualified properties and locating them on a map. You will perform that task with Business Analyst's Store Setup tool. You will then use a variety of trade area and reporting tools to explore the locations of these potential sites relative to concentrations of attractive customers, centers of retail traffic, and competing home centers. In chapter 6 you will use this information to make your site selection decision.

Run Business Analyst Desktop; geocode available properties locations

1. Click Start, Programs, ArcGIS, Business Analyst, BusinessAnalyst.mxd to run ArcMap, load the Business Analyst Extension, and then load the default Business Analyst map.

2. Click OK when the Update Spatial Reference dialog box appears, then close the Business Analyst Assistant window on the right of the screen.

3. Click File, click Open. Navigate to C:\My Output Data\Projects\LITGL Minneapolis St Paul\CustomData\ChapterFiles\Chapter4\LITGLFirstStore.mxd. Click the map file to open it.

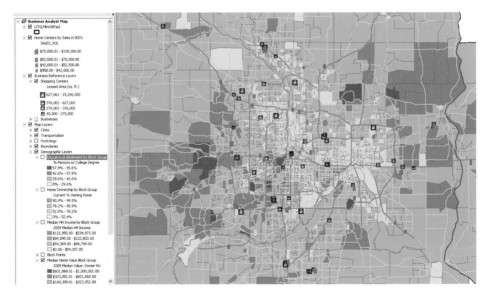

4. Click the Source tab at the bottom of the Table of Contents.

In Source view, the layers are organized by their source files and their locations. Note that three of the thematic layers are based on BGIncEdOwnExp.shp, a shapefile identical to the one you created in chapter 3. The fourth, Median Home Value, is based on the original Business Analyst block group dataset. Turn layers on and off to familiarize yourself with the maps. When you are done, turn off the Home Centers by Sales Volume and Shopping Centers layers.

5. Click the Display tab at the bottom of the Table of Contents to return to Display view.

    The thematic maps you have created provide the backdrop for the trade area and site selection analysis you will perform in this chapter. A commercial real estate agent has provided you with a list of available retail properties in Microsoft Excel format. To integrate them into the project, you will geocode them and adjust their symbolization to reflect Janice's and Steven's site preferences.

6. Click the drop-down arrow on the Business Analyst toolbar, click Store Setup to initiate the Store Setup wizard. In the first window, select Create New Store layer, click Next. In the resulting window, select Tabular data, click Next. In the resulting window, select In a file on my computer, click Next. Click the Open file icon, navigate to C:\My Output Data\Projects\LITGL Minneapolis St Paul\CustomData\ChapterFiles\ Chapter4\AvailableProperties.xls. Double click this file to view its contents, select the StoreLayer$ table, click Add. Click Next.

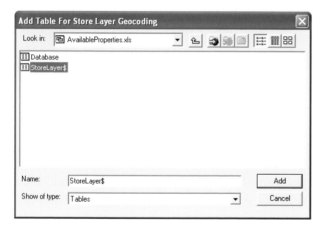

7. Click Next to view the Input Fields window. Review the settings and accept the default values (Address, City, State, and ZIP) by clicking Next. In the resulting window, accept the default "Name" and "ID" field settings by clicking Next. In the final window, enter **Available Properties** in the name field. Click Finish.

    Business Analyst geocodes the store addresses in the table, displays them on the map and adds a layer to the Table of Contents which uses a single symbol to depict the stores. As some of the locations fall outside the ideal size range selected by Janice and Steve, you may wish to symbolize them by size relative to that range. To do so, you use the Symbolize tab of the Layer Properties window as you did in chapter 3.

8. Open the Layer Properties window for the Available Properties layer. Click the Symbology tab. Select Quantities, then Graduated symbols in the Show field. Select Size(SqFt) in the Fields Value field box and *none* in the Normalization box.

9. Click Classify and specify three classes in a Manual classification scheme. Enter **39,000**, **60,000**, and **80,000** as break values. Click OK.

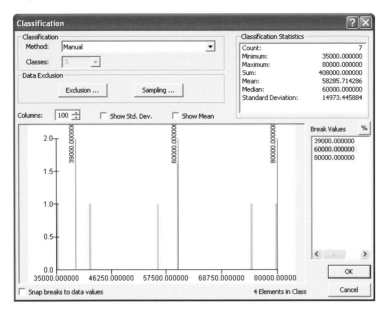

10. Click the Template button and select Fire Red triangles as the symbol. Set 25 to 15 as the Symbol size range. Format the labels to display numbers with no decimals and with thousands separators. If necessary, click Symbols, then Flip Symbols to move the largest symbol to the top of the legend. If necessary, click Range, Reverse Sorting to display the highest values at the top of the legend. Select the Show class ranges using feature values option. The Symbology window should resemble the one below. Click OK.

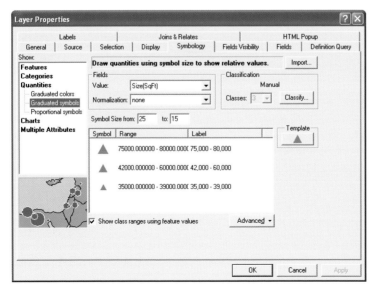

Business Analyst applies the settings to the map and Table of Contents. Your map should resemble the one below, though the thematic layer displayed might differ. The available sites are displayed in three size classes. The middle class includes sites within the desired size range, while the upper and lower classes include sites that fall below or exceed that range. These sites will not be eliminated from further consideration, but their size variance will affect the final decision.

11. To preserve your work, save your map file as **LITGLFirstStore2.mxd** to C:\My Output Data\Projects\LITGL Minneapolis St Paul\CustomData\ChapterFiles\Chapter4\.

You will now use the customer prospecting and trade area functions of Business Analyst Desktop to broaden your understanding of these sites.

Use Customer Prospecting and Grids to compare available sites to customer profile characteristics

Customer prospecting analysis in Business Analyst identifies the geographic units in a region that meet criteria established by the user. You will use it to identify the most attractive block groups in the Twin Cities area relative to the characteristics in the green-customer profile. Recall that the characteristics of this profile include above-average levels of income, education, home ownership, and home value. Thus, concentrations of households with above-average values on all four of these measures would be quite attractive to Janice and Steven. You will use Customer Prospecting Analysis to identify block groups which match this profile.

1. Click the drop-down arrow in the Business Analyst toolbar, click Analysis. Select Create New Analysis, click Next. Select Customer Prospecting, click Next. Select Median Home Value by Block Group as the level of geography to prospect, click Next. Select

Use feature from a layer and select the LITGLMinnStPaulSA layer, click Next. Change the entry method to Enter values manually, click Next.

The next window contains a box listing the attributes available in the layer. You will use this list to designate the four criteria for selecting block groups. Confirm that the Match all criteria (AND) option is selected at the top right of the dialog box.

2.   Click the attribute CY Median HH Income in the CY Household Income category, click Show in the Field Statistics button at the bottom right of the window to calculate and display descriptive statistics for this attribute.

The mean value in the graphic above is 75,090, though it will vary depending on the block groups included in the current extent. Be sure you understand what this value represents. It is the mean of the values for each block group and those values are medians. Thus it is *not* accurate to say the mean household income is $75,090. Rather this number is the mean of median block group values. You wish to select block groups with values at or above that figure.

3.   Double click the CY Median HH Income attribute and enter the value you just calculated into the Floor box.

The Analysis Wizard window should look like this.

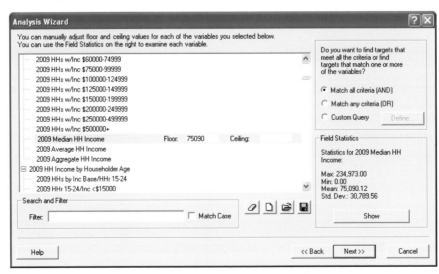

4. Repeat this procedure to select the attribute CY Median Value: Owner HU (median home value) in the CY Home Value category, calculate its average (198,475), and enter this value as the Floor criterion. Repeat for the attribute CY Pop 25+ by Educ: Bach Deg in the CY Pop 25+ by Educ Attainment category (233). Repeat again for the attribute CY Owner Occupied HU in the 2009 Demographic Overview category (415).

The Analysis Wizard window should resemble the one below. When applied to the data layer, it will select those block groups which meet all four of these criteria.

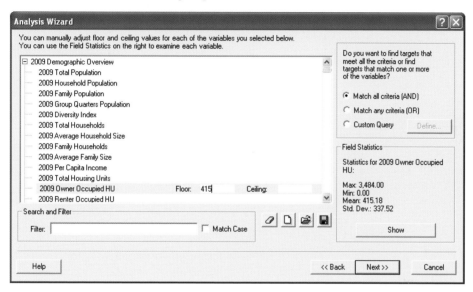

5. Click Next. Enter **LITGL Customer Prospecting** in the name box. Click Finish.

Business Analyst applies the selection criteria you designated, identifies the block groups that meet all four criteria, creates a layer containing these features, adds it to the Table of Contents, and displays it on the map, which should resemble the one below. You may use this map to assess the distribution of Available Properties relative to these concentrations of attractive customers visually.

 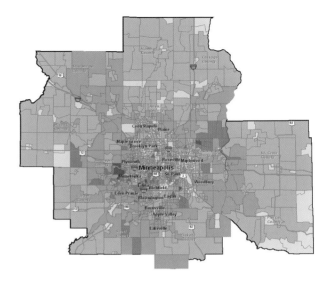

You also wish to compare the locations of these sites to home-related expenditures by household across the study area. To do so, you will use Business Analyst's Grid Trade Area function.

6. Turn off the LITGL Customer Prospecting and Home Centers by Sales Volume layers. Right-click the Available Properties layer, click Zoom to Layer to zoom the map to an extent that includes the seven available sites.

7. Click the drop-down arrow in the Business Analyst toolbar, click Trade Area. Select Create New Trade Area, click Next. Select No Customer Data Required, click Next. Select Grids, click Next. Select Use Current Extent, click Next. Enter 2 miles as the grid size and distance, click Next. In the Select Layer box, select Standard Business Analyst Data (Generalized). In the Symbolization field box select Shelter: Tot, which you will find in the Expenditures data near the bottom of the attribute list, click Next.

8. In the resulting window, add four additional relevant attributes (CY Total Households, CY Owner Occupied HUs, CY Median HH Income, and CY Med Value Owner HU) to the list of Selected variables. The window should resemble this. Click Next.

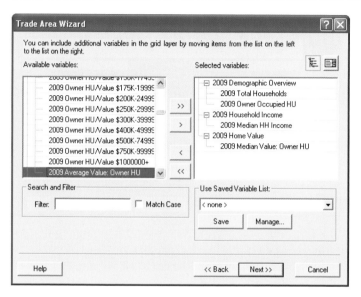

9. Enter **Total Shelter Expenditures Grid** in the Trade Area name box, click Finish.

In a process that might take several minutes, Business Analyst divides the current extent of the map into a grid of cells with an area of four square miles (2×2 mile cells), aggregates home-related expenditures for each cell, adds the grid to the Table of Contents, symbolizes each cell, and produces a map similar to the one below. As the color scheme is ascending values of red, this is also known as a heat map, with the hotter (darker red) cells representing higher levels of expenditures. Use the heat map to assess visually the location of available properties to areas with high levels of home related expenditures.

You may need to change the color of the Available Properties symbol and move this layer to a position above the grid layer to make this comparison. The map below includes these changes and illustrates the appropriate configuration of the Table of Contents.

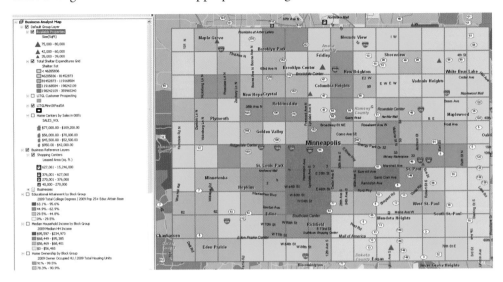

You may view the values of individual cells in the grid by selecting the Identity tool in the Tools toolbar and clicking within the boundaries of the target cell. This allows you to view the values of all the attributes you specified for each grid. Use this tool to view the attributes in the vicinity of each store and compare them to the cells with the highest levels of home related expenditures, i.e., those with the darkest red color.

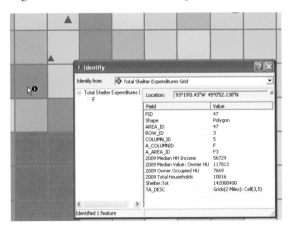

10. When you are finished, click the Select Elements tool ➤ to turn off the Identity tool. Turn off the Total Shelter Expenditures Grid layer.

Create Threshold Ring trade areas for available sites

While the heat map approach displays the distribution of consumer attributes across the area of interest, threshold rings display similar information for specific locations. You will use this function to explore the distribution of expenditure data relative to the available properties. Threshold rings display the geographic concentration of a designated attribute around a specific location. Commonly designated attributes include market size measures such as households, population, or, in this case, consumer expenditures. Thus, threshold rings can answer questions such as "How large must a store's trade area be to include 50,000 households? 75,000 people? $20,000,000 in relevant product purchases?" Further, by creating multiple rings at various levels of these attributes, the user can assess the resulting concentration at several relevant thresholds. You will use it to assess the concentration of total shelter expenditures around each of the available sites.

Based on their cost, revenue, and market share projections, Janice and Steven believe that the first Living in the Green Lane store must serve a market with at least $75,000,000 to $125,000,000 in total shelter expenditures. (While this category of expenditures is broader than LITGL's product lines, it does include them all and is a good indicator of overall consumer spending on shelter.) You will use these values to create threshold rings around available properties. You will then examine these trade areas relative to each other and to competing home centers.

1. Confirm that the grid and customer prospecting layers are turned off. Open the Layer Properties window of the Available Properties layer and, if necessary, change the color of its symbol to bright red or orange, then close the window. Right-click this layer and click Zoom to layer.

2. Click the drop-down arrow in the Business Analyst toolbar, click Trade Area. Select Create New Trade Area, click Next. Select No Customer Data Required, click Next. Select Threshold Trade Areas, click Next. Select Available Properties as the store layer and ID as the ID field. Select the All Stores option, click Next.

3. In the resulting window, select Median Home Value by Block Group as the Threshold layer and Shelter: Tot as the field to aggregate, click Next. (Note: Should you receive a warning message "This aggregation method for the selected field must be set to 'SUM', Please select another field" in this process, click OK, but do *not* select another aggregation field.) Select the Use Standard Threshold Rings option, then click Next.

4. In the resulting window, select 2 as the number of rings you wish to create and enter **75,000,000** as the value for the inner ring and **125,000,000** as the value for the outer. The window should resemble the one below. Click Next.

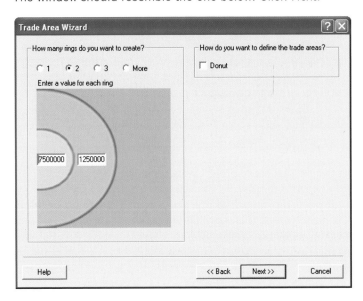

5. Enter **Total Shelter Expenditures Threshold Rings** as the Trade Area name. Confirm that the Create Reports option is *not selected*. Click Finish.

In a process that might take several minutes, Business Analyst performs the necessary calculations, creates the resulting threshold rings, adds them to the Table of Contents, and displays them on the map, which should resemble to one below. (Note: The major roadways may not be visible due to scale dependency. If that is the case, turn on the Major Highways (Regional) layer in the Streets and Highways group layer in the Table of Contents. You should also turn on the Shopping Centers and Home Centers by Sales Volume layers to match the map on next page.)

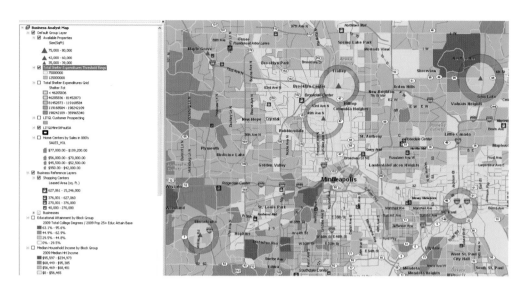

Compare the rings of the available properties. The smaller the rings, the more compact the concentration of home-related expenditures around the location. Compare the locations to those of competing home centers and shopping centers as well. Shopping centers generally attract retail traffic while home centers are LITGL's most direct competitors.

The Customer Prospecting, Grid, and Threshold Rings functions of Business Analyst Desktop allow you to compare available sites to customer characteristics. The Threshold Rings approach also provides useful competitive comparisons. Now you wish to explore the competitive environment more fully by creating Huff Equal Probability Trade Areas around existing home centers.

Use Huff Equal Probability Trade Areas and Locator Reports to compare the competitive environments of available sites

In Business Analyst Desktop two trade area models—Equal Competition (Thiessen) polygons and Huff Equal Probability Trade Areas—are particularly useful for exploring a company's competitive landscape. The Equal Competition model creates polygons around each store. The boundaries of each polygon are composed of points equidistant from that store and its nearest competitor. Thus, if linear distance were the only decision factor for consumers, each store would attract the consumers within its polygon.

The Huff Equal Probability Trade Areas approach is similar, but allows users to include factors other than linear distance in estimating the probability that consumers will shop at a given store. Janice and Steven believe that store sales volume as measured in annual sales is as important as linear distance when consumers choose between competing home centers. For this reason, you will use Huff Equal Probability Trade Areas approach to create trade areas around selected competing home centers based on both linear distance and sales. The result will be trade areas around each store in which consumers are more likely to shop at that store than at competing stores. Boundaries between polygons represent areas where consumers are equally likely to shop at either store, hence the name of this approach. You will select stores for the analysis based on sales volume and proximity to available properties.

1. Right-click the Home Centers by Sales in 000's layer, click Selection, then click Make This the Only Selectable Layer. This allows you to select the specific home centers you wish to include in the Huff analysis.

2. In the Tools toolbar, click the button for the Select Features tool . Press the Shift key, then click and drag the cursor to form a rectangle around a large home center near one of the available properties. The feature is highlighted to indicate its selected status. With the Shift key depressed, repeat this process for another home center. Note that it is added to the selection while the first one is also retained. (If you perform this task without the Shift key, the second selection would replace the first.) Repeat this process to select the home centers that you believe will pose the greatest competitive challenge to the available properties you are evaluating. This should include large home centers, conveniently located on major roads. You should select about 15–20 home centers. Your screen will resemble the one below, though your selection may vary. (Note: In this layer, the labels for the Detailed Regional Highways layer is turned off. If these labels obscure your screen, you may turn them off in the Labels tab of the Layer Properties window for this layer.)

3. Click the drop-down arrow in the Business Analyst toolbar, click Trade Area. Select Create New Trade Area, click Next. Select No Customer Data Required, click Next. Select Huff Equal Probability Trade Areas, click Next. Select Home Centers by Sales in 000's as the store layer and LOCNUM as the ID field. Select the Selected Stores option, click Next.

4. In the resulting window, select the Enter parameters manually option, click the + button to the right of the window to add a new parameter. Select SALES_VOL as the additional variable and enter a coefficient of 1.5.

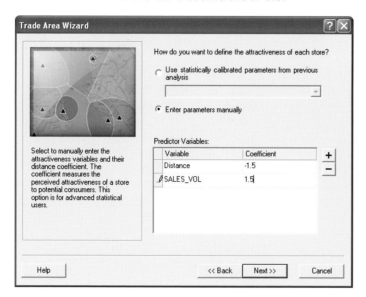

The window should resemble this. Consider the entries for a moment. The coefficient for distance is negative, indicating an inverse relationship with store attractiveness. That is, the further a store is from a consumer, the less attractive it is. The coefficient for SALES_ VOL is positive indicating a direct relationship. The higher a store's sales volume, the more attractive it is to consumers. The relative importance of these factors is expressed in the magnitude of the values. You have set them as equal, but you could easily adjust the values should you deem distance to be more important than sales volume, or vice versa. In addition, you could include more factors with variable weights in the analysis. This analysis requires only the two you have entered.

5. Click Next. Enter **Huff Equal Probability Trade Areas** as the Trade Area name. Confirm that the Create reports option is *not* selected. Click Finish.

Business Analyst Desktop calculates the boundaries of the trade areas and displays them on a map resembling the one on page 104. Each polygon represents the customers most likely to shop at the home center it surrounds based on the location of that facility and its sales volume. Notice that the largest polygons are on the periphery of the study area, where no home centers are located. These areas appear to present opportunities for new competitors. However, comparison of this map with the grid and consumer profiling maps you created earlier indicates that this is not the case.

Move the Available Properties layers to a position above the Huff Equal Probability Trade Areas layer so they will display over this layer. Examine the location of sites in the Available Properties layer relative to this map. Generally, locations near the boundaries of trade areas in the Equal Probability map indicate favorable opportunities, while those in the central regions of polygons indicate highly competitive environments. Note that several available properties are located near such boundaries.

While this map portrays the competitive environment visually, you wish to learn more about the home centers competing with each available site as well as the shopping centers which attract customers to their vicinity. You will design Locator Reports to do so.

6. Click the Select Elements button ▶ in the Tools toolbar to turn off the Select Features tool. Click the drop-down arrow on the Business Analyst toolbar, click Reports. In the Reports Wizard, click Run Reports, click Next. In the resulting window, click Run point and ranking based reports, click Next. In the resulting window, click Locator Report, click Next.

7. In the resulting window, select Available Properties as the store layer, ID as the ID field, and Name as the Name field. Select All Sites. Click Next.

8. In the resulting window, select Home Centers by Sales in 000's as the business points layer, select the Limit report to closest option, enter **10** in the left box, click Next.

9. In the resulting window, select Drive Time as the distance calculation method, select the Add distance back to business points layer option, check the Create new distance field option, enter **Miles** as the field name then click Next.

10. In the resulting window, select the Landscape option and specify the attributes you wish to include in the report by matching your entries to those in the window below. Click Next.

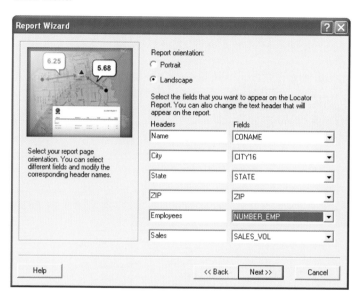

11. In the resulting window, select the Generate Report and View report options, enter **Home Centers Locator Report** in the Report name box and **Competing Home Centers by Site** in the optional Report title box, click Finish.

Business Analyst identifies the competitors within a 10-mile linear distance of each available site and generates the report based on your settings. For each site, the report identifies competitors within 10 miles and displays the number of employees, sales in thousands of dollars, and the direction and distance in drive time for each. The report should resemble the one below.

Home Centers Locator Report

Name	City	State	ZIP	Employees	Sales	Dir.	Minutes
1							
Garrison Bldg							
LOWE'S	WEST ST PAUL	MN	55118	0	2450	SE	6.43
MENARDS	MAPLEWOOD	MN	55109	120	42000	NE	6.47
HOME DEPOT	INVER GROVE HTS	MN	55077	0	56000	SE	6.58
MENARDS	ST PAUL	MN	55104	200	70000	NW	6.87
HOME DEPOT	ST PAUL	MN	55125	140	49000	SE	8.56

Review the drive-time numbers for competing home centers for each site. For reasons discussed below, you will define LITGL's potential market area with 3- and 6-minute drive-time polygons. If competing home centers have similar trade areas, those with drive-time values of 10 minutes or less will overlap the trade area of the available site to some extent.

Shopping centers also affect the competitive environment of each site by drawing retail traffic to the area. Thus, you wish to determine how many shopping centers lie within the 6-minute trade areas of each available site. You will design a Locator Report for this purpose as well.

12. Minimize the Locator Report window for later use. Repeat steps 4 to 9 above, with these changes. In step 6, designate Shopping Centers as the business points layer and change the distance from **10** miles to **6**. In step 8, select the Portrait option and match your settings to those in the window below:

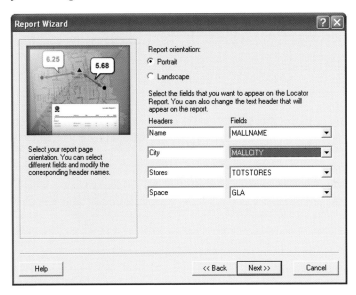

13. In step 9, enter **Shopping Centers Locator Report** in the Report name box and **Nearby Shopping Centers by Site** in the Report title box, click Finish. The resulting report should resemble the one below.

Name	City	Stores	Space	Dir.	Minutes
1					
Garrison Bldg					
Sun Ray Shopping Center	St. Paul	33	287385	NE	4.63
Midway Marketplace	St. Paul	14	487196	NW	4.73
Midway Shopping Center	St. Paul	40	293732	NW	4.88
Signal Hills Shopping Center	West St. Paul	25	225000	SE	5.52
Har-Mar Mall	Roseville	36	432614	NW	9.10
Total number of business points:	5				
2					
Reynolds Bldg					
Maplewood Mall	St. Paul	142	922289	SE	7.01
Total number of business points:	1				

Review the Locator Reports and the Huff Equal Probability Trade Areas layer you have created. In the next chapter, you will use them to assess the competitive environment of each of the available sites.

14. Minimize the Locator Report window for later use. Turn off the Huff Equal Probability Trade Areas layer.

15. Save your map file.

This completes the tasks in this chapter. Beginning with a simple Microsoft Excel table of available locations you have geocoded and mapped those locations, used Business Analyst Desktop trade area tools to learn more about them, and generated Locator Reports to gain insight into their competitive situation. In chapter 5, you will build on this analysis by integrating this information with data on specific trade areas to select the best site and design map documents to support your decision.

Chapter 5

Defining trade areas, generating reports, selecting best site

In chapter 4, you used the trade area functions of ESRI Business Analyst Desktop to generate a significant body of information on consumer and competitive factors in the area and their spatial relationship with shopping centers, competing home centers, and the sites available for the first LITGL store. Further, you used the Threshold Rings method to create trade areas around the available sites. While this method captures concentration of sales for each site, Janice and Steven do not deem it to be the most appropriate trade area method for Living in the Green Lane.

To see why, toggle the Shopping Centers and Home Centers by Sales Volume layers and notice how these features cluster around major highways. This emphasizes the importance of transportation infrastructure to these retail operations. Further, Janice and Steven wish to maximize the convenience and minimize the transportation resources required to visit Living in the Green Lane's store. As you learned in chapter 2, the simple ring approach uses linear distance as a measure of convenience. As you also learned, drive-time polygons are a more appropriate approach if natural barriers or transportation infrastructure are significant factors, as they are here. For these reasons, they wish to use drive-time polygons to create trade areas for the available properties.

In addition, Janice and Steven wish to define compact trade areas. While they believe that LITGL's approach creates benefits for which consumers would be willing to drive some distance, they also wish to develop the concept as a local, neighborhood resource with a concentrated consumer base. Thus, they wish to use 3- and 6-minute drive times as the thresholds for drive-time polygons. Using these settings, you will create drive-time polygon trade areas for the available sites and generate reports on their comparative characteristics.

Run Business Analyst Desktop; Load map

1. Click Start, Programs, ArcGIS, Business Analyst, BusinessAnalyst.mxd to run ArcMap, load the Business Analyst Extension, and then load the default Business Analyst map.

2. Click OK when the Update Spatial Reference dialog box appears, then close the Business Analyst Assistant window on the right side of the screen.

3. Click File, click Open. Navigate to C:\My Output Data\Projects\LITGL Minneapolis St Paul\CustomData\ChapterFiles\Chapter5\LITGLFirstStore.mxd. Click the map file to open it.

This map is very similar to the one you designed in chapter 4. You will use it to continue your analysis. Note the Customer Prospecting, Huff Equal Probability, and Total Shelter Threshold Rings layers just below the Businesses layer. You will use these layers later in this chapter. Note as well that many of the layers in the map are turned off to make the Available Properties layer and the trade areas you will create around it more visible.

Create drive-time trade areas for available sites and generate reports

1. Right-click the Available Properties layer, click Zoom to Layer. Open the Layer Properties window for this layer. Click the Labels tab at the top of the window. Select the Label Features in this Layer option, designate Name in the Label Field box, select Times New Roman as the font and 12 as the font size. Select the Bold option. The Label Settings window should look like the one below. Click OK to close the window, apply the settings, and add the labels to the map.

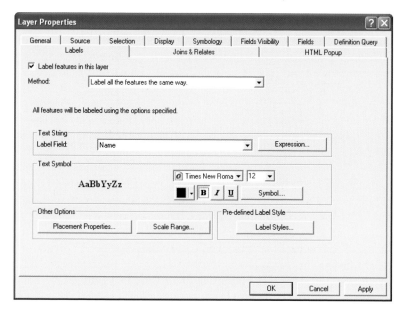

2. Click the drop-down arrow in the Business Analyst toolbar, then click Trade Area. Select Create New Trade Area, then click Next. Select No Customer Data Required, then click Next. Select Drive Time Polygons, then click Next. Select Available Properties as the store layer and ID as the ID field. This designation will make the reports easier to match with the sites on the map. Select the All Stores option, then click Next.

3. In the resulting window, select 2 as the number of drive-time trade areas, select Minutes in the distance units box, then enter **3** as the value for the inner polygon and **6** as the value for the outer polygon. Confirm that Network Analyst is selected in the Drive Time Polygons box. Confirm that the window looks like the one below. Click Next.

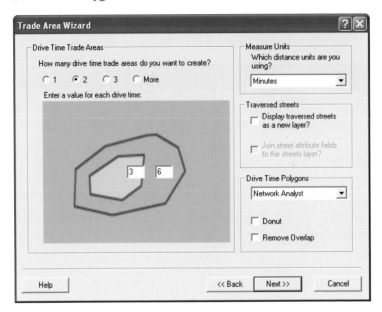

4. In the resulting window, enter **Available Properties Drive Time Polygons** as the name of the new Trade Area. Select the Create Reports option. (Note: Though this option is available for all trade area functions in Business Analyst Desktop, this is the first time you have selected it. This option allows you to select from a comprehensive range of reports with information on each trade area. If you wish, you may use the Trade Area function to create similar reports for any of the trade areas you have already created.)

5. Click Next, select the For Individual Features option, click Next to open the Report Templates window. (Note: If your system generates an error message at this stage, click on the Options button below the templates box on the left side of the screen. In the resulting window, use the drop-down box at the top of the window to select the Standard BA Data (Generalized) as the layer to summarize, then click OK. When given the option to set this as the default layer for reports, click Yes. You will return to the Reports box with the available report templates listed in the box on the left.)

6. The Report Templates window lists the available report templates in the box on the left. To select a report, simply move it from the box on the left to the box on the right with the arrow keys between the two boxes. Click the Market Profile Report to select it, then click the right arrow between the boxes to move it to the box on the right and order this report. Click Next.

The content of the Business Analyst report templates is described on pages 86 and 87 in the Introduction to Part III. You have selected the Market Profile Report because it contains data on all the characteristics of the green consumer profile. As you can see from the description above, other reports would be more appropriate for different segmentation schemes. Because reports can be very lengthy, it is useful to use the most relevant report(s) as the initial screening tool and use other reports to gather more detailed information on the sites that remain in contention as the screening process proceeds.

7.   In the final window, select only the View Reports on Screen option. (You can export a report to a data file in several formats after your initial reading.) Click Finish.

Business Analyst Desktop creates 3- and 6-minute drive-time trade area polygons for each available site, calculates the values for the report, adds the trade areas to the Table of Contents, displays them on the map and opens a window to display the Market Profile Report for each of the available sites. Your map should resemble the one below.

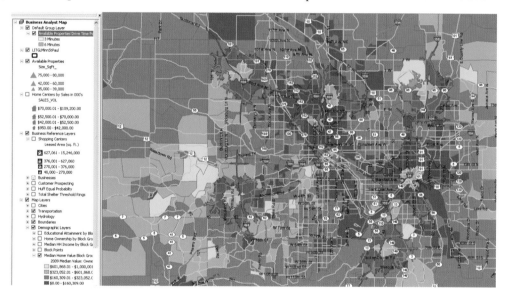

The report will display in a new window. You may view it there and export it to a format of your choice to have it available for future reference. This report will be your first source of information in selecting a site for LITGL's store.

8.   Save the map to **LITGLFirstStore2.mxd** in C:\My Output Data\Projects\LITGL Minneapolis St Paul\CustomData\ChapterFiles\Chapter5\ to preserve your work.

Use maps, Market Profile Report, and Locator Reports to screen available sites

There is no single method for integrating the range of data and maps you have generated into your site selection decision. The following approach focuses on trade area population characteristics, home-related expenditure data, and competitive considerations to identify the most attractive sites. It then uses additional Business Analyst Desktop reports to incorporate more information into the analysis to reach a conclusion and recommendation. You will begin by comparing the population characteristics of the available sites with the characteristics of the green consumer profile.

1.   Remember that green consumers tend to exhibit above-average levels of education, income, and home ownership. The Market Profile Report contains data on each of these characteristics for the drive-time polygon trade areas you created for each available site. You will use this data to identify the most attractive locations.

   Use the data from the Market Profile Report to record data on trade area characteristics in the table below. Enter current year values for the 6-minute polygon for each site.

Trade area	Total households	Median HH income	% owning home	% college degree *	Median home value
Garrison					
Reynolds					
Mayer					
Steiers					
Hall					
Carter					
Tucker					
* % college degree is the sum of the associate, bachelor, and graduate/professional degree percentages					

Table 5.1 Population characteristics of available trade areas

Compare the values for the available sites. Which sites most closely match the characteristics of the green consumer profile?

2. Use the data from the Market Profile Report to record data on home-related purchases for each available site. Enter values for the 6-minute polygon for each site.

Trade area	HH furnishings expenditures			Shelter expenditures		
	Total $Mil	per HH	Index	Total $Mil	per HH	Index
Garrison						
Reynolds						
Mayer						
Steiers						
Hall						
Carter						
Tucker						

Table 5.2 Home-related expenditures in available trade areas

Total expenditures reports the total level of spending on this category in the trade area. Per HH (household) reports the average spending per household. The Index compares average household spending in each category to the national average. Thus, an index of 150 means that average spending on the category in the trade area is 50 percent higher than the national average, while a value of 75 means it is 25 percent lower.

Which available properties display the most attractive spending patterns?

3. Toggle the Home Centers by Sales Volume, Shopping Centers, and Huff Equal Probability Trade Areas layers one at a time to compare them. Note the locations of available properties relative to these features and the transportation infrastructure of the area.

How are available sites positioned relative to shopping centers which attract retail traffic and transportation infrastructure which facilitates it?

How are available sites positioned relative to competing home centers and the boundaries of Huff equal probability trade areas?

4. Use the data from the Home Center and Shopping Center Locator Reports you generated in chapter 4 to record competitive environment data for each site.

Trade area	Attractors Shopping centers within 6 minutes		Competitors Home centers within 10 minutes	
	Number of centers	Number of stores	Number of competitors	Sales $US Mil
Garrison				
Reynolds				
Mayer				
Steiers				
Hall				
Carter				
Tucker				

Table 5.3 Shopping center and home center drive-time distances from available sites

Based on the maps and Locator Reports, which sites have the most favorable competitive environment? The most unfavorable? Explain.

Recall that Janice and Steven wish to purchase a retail facility with 40,000 to 60,000 square feet of floor space with four to five parking spaces per 1,000 square feet of retail space. They are willing to consider larger or smaller facilities if they serve highly desirable trade areas.

5. Right-click the Available Properties layer, then click Open Attribute Table to view the layer's attributes. Compare the attributes of each site with Janice and Steven's criteria.

Which sites meet the selection criteria?

Based on population characteristics, home-related expenditures, competitive factors and site characteristics, which two sites are the most favorable? Explain why.

Use Detailed Reports to select the site for the first Living in the Green Lane store

Janice and Steven agree with your conclusion that the Steiers and Carter buildings appear to be the most favorable sites available. They ask you to explore them further and recommend one of the sites. In that process, they ask you to address four questions.

- The Carter site's median age is higher than the Steiers site and its average household size lower. Do these factors reflect differences in the age, employment, and family composition of the two sites? If so, how might these factors affect Living in the Green Lane's sales in these trade areas?
- Are the favorable income, home ownership, and home value characteristics of these sites projected to continue through the next five years? Which site will have the greatest growth in population? In households?
- Both sites have high spending indices for general home-related expenditures. Is this also true for more detailed categories of expenditures in this area? What implications do these indexes have for Living in the Green Lane's market potential?
- Do the age and household size differences between these two sites reflect underlying differences in lifestyle segments as well? If so, how might these differences affect Living in the Green Lane's marketing strategy in these trade areas?

Some of the information you need to answer these questions is available in the Market Profile Report you have created. However, you must create additional reports to answer some of the other questions. Specifically, the Comprehensive Trend Profile Report will provide relevant information for the second question, the Retail Expenditure Report for the third and the Tapestry Segmentation Area Profile Report for the fourth. As you need these reports only for the Carter and Steiers sites, you will select those two trade areas and order the necessary reports for them.

1. To open the layer's attribute table, right-click the Available Properties Drive Time Trade Areas layer, then click Open attribute table.

2. Click in the small gray boxes to the left of the features with Area_ID Steiers Bldg_1, Steiers Bldg_2, Carter Bldg _1, Carter Bldg _2 to select them.

   These are the 3- and 6-minute trade areas for the Steiers and Carter sites. When you select them, they will be highlighted in the attribute table and the corresponding trade areas will be highlighted on the map as well. Your map should resemble the one on next page.

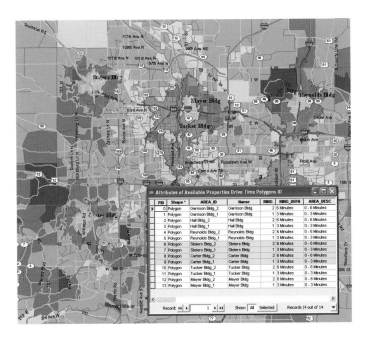

3. Click the drop-down arrow in the Business Analyst toolbar, then click Reports to open the Report Wizard. Click Run reports, then click Next. Click Run reports for a single layer, then click Next.

4. In the resulting window, select Available Properties Drive Time Polygons as the boundary layer and select the Use selected features only option. The window should resemble this. When it does, click Next.

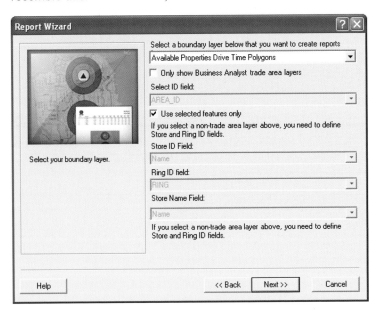

5. In the resulting window, select the For Individual Features option, then click Next. In the next window, remove the Market Profile Report from the selected window and then move the Comprehensive Trend Report, Retail Expenditure Report, and Tapestry Segmentation Area Profile from the box on the left to the box on the right using the arrow keys between the boxes. The window should resemble this. When it does, click Next.

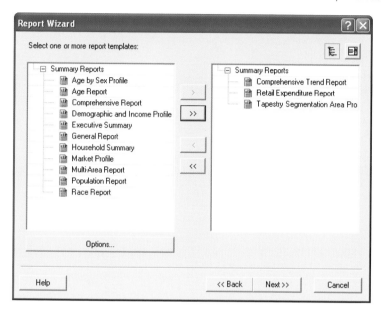

6. In the final window, select the View reports on screen and Create single report file options. Click Finish.

Business Analyst calculates the necessary values, prepares the reports, combines them into a single document, and displays them in the Report Window. You may use the Export function in the upper left corner of the window to export the combined reports to a format and location of your choice. The reports are now ready for you to use in answering Janice and Steven's questions.

The Carter site's median age is higher than the Steiers site and its average household size is lower. Do these factors reflect differences in the age, employment, and family composition of the two sites? If so, how might these factors affect Living in the Green Lane's sales in these trade areas?

The Market Profile Report provides the information you need to answer this question. Begin by completing the following table for the 6-minute drive-time trade areas of the two sites. The first two rows are the statistics Janice and Steven mention in this question.

Population characteristic	Steiers Building 6-minute drive time	Carter Building 6-minute drive time
Median age	36.4	44.2
Average household size	2.71	2.48
Not in labor force, 2000		
Households with related Children, 2000		
Households with persons 65+, 2000		
Median year moved into present house, 2000		
Median year structure built, 2000		

Table 5.4 Population characteristics of Steiers and Carter trade areas

Use this data to answer the first question here.

Are the favorable income, home ownership, and home value characteristics of these sites projected to continue through the next five years? Which site will have the greatest growth in population? In households?

The Market Profile Report and the Comprehensive Trend Report provide the data necessary to answer this question. Review these reports and answer the second question here.

Both sites have high-spending indexes for general home-related expenditures. Is this also true for more detailed categories of expenditures in this area? What implications do these indexes have for Living in the Green Lane's market potential?

The Market Profile Report includes only two general expenditure categories for Household Furnishings and Equipment and Shelter. As referenced in the question, the Carter and Steiers trade areas have high-spending indexes for both categories. The Retail Expenditures Report includes these categories but also several subcategories within each. It also reports the spending index of each trade area for these subcategories. Review these values for the Steiers Building 6-minute trade area and the Carter Building 6-minute trade area. High-spending indexes across the subcategories would obviously be favorable to LITGL. Use this information to answer the third question here.

Do the age and household size differences between these two sites reflect underlying differences in lifestyle segments as well? If so, how might these differences affect Living in the Green Lane's marketing strategy in these trade areas?

Business Analyst Desktop uses the Tapestry Segmentation Neighborhood Segmentation system to classify block groups in the United States into 65 distinct lifestyle segments. Lifestyle segments differ in demographics, values, housing characteristics, activities, and purchasing patterns. Those differences can be crucial in selecting appropriate sites, but also in crafting marketing strategies to serve customers well.

The variations between the Steiers and Carter sites in age, employment, household size, and housing patterns suggest that each may contain different Tapestry Segmentation segments. The Tapestry Segmentation Area Profile Report provides the data to make that determination. In that report, Tapestry Segmentation Segments are organized by Life Mode and Urbanization groups. Segments within the same Life Mode or Urbanization group share some common characteristics. Within this structure, the report lists the number of households in each Tapestry Segmentation segment for each of the two trade areas. It also provides a segment index, which compares the percentage of trade-area households in the segment to the corresponding percentage of national households in that segment. Higher values for this index indicate greater concentrations of that segment in the trade area.

Use the data in the Tapestry Segmentation Area Profile report to identify the six most common Tapestry Segmentation segments in each trade area, recording the percentage of trade area households and index of each.

Steiers trade area 4_2			Carter trade area 6_2		
Segment number and name	% of HHs	Index	Segment number and name	42.1	Index
Total		xx	Total		xx

Table 5.5 Major Tapestry Segmentation segments by trade area

You will use expanded descriptions of Tapestry Segmentation segments in later chapters. To answer the current question, compare the general characteristics of the dominant segments in each trade area by using the Tapestry Segmentation Summary Table included in the Business Analyst Desktop documentation.

7.   Navigate to C:\Program Files\ArcGIS\BusinessAnalyst\Documentation\USA_ESRI_ Tapestry_Summary_Tables.pdf and double-click the file to open the Tapestry Segmentation Summary Table.

This table provides an overview of the core characteristics of each Tapestry Segmentation segment. Review the entries for the dominant segments in each trade area. The more closely these values reflect those of the green consumer profile, the more attractive the segment is for Living in the Green Lane.

Using the data from this report and the two tables, answer the fourth question here:

> **Based on your earlier analysis and your answers to these four questions, which location—Steiers or Carter—will you recommend as the site of Living in the Green Lane's first store? Explain your reasoning.**

Use Layout View to design a map document to support your conclusions and recommendation

Janice and Steven have decided upon the Steiers location as the site for their first store. They wish to include two maps in their business plan to communicate their decision process and support the recommendation. The first map will display the concentration of the most at- tractive customers in the Twin Cities area relative to available locations. The second will dis- play the Steiers trade area and its competitive environment. You will use the Layout features of Business Analyst Desktop to create these maps.

Begin by reviewing the layers of your map and determining which you wish to display in the map document. For the first map, you will use the Educational Attainment by Block Group layer as the basemap, with the major highways, Available Properties, Total Shelter Expendi- tures Threshold Rings, Home Centers by Sales in 000's, and the LITGL Customer Prospect- ing layers displayed above it. Turn those layers on and the remaining layers off. Be sure that the labels for the Available Properties layer are displayed and turn them on if they are not.

1.   Right-click the Total Shelter Expenditures Threshold Rings layer, then click Zoom to layer to set the extent of the map to include all the features of this layer. Open the Layer Properties window for this layer, click the Symbology tab and, in the Labels column directly under the Color Ramp box, replace 75000000 with **$75 Million** and 125000000 with **$125 Million**. Click OK.

     Note that the Customer Prospecting layer obscures the features beneath it. You will increase the transparency of this layer to correct this.

2.   Open the Layer Properties window for the Customer Prospecting layer. Click the Display tab, enter **20** in the Transparent box to set the transparency for this layer to 20 percent. If necessary, turn off the drive-time trade areas layer and adjust the symbol size of the Available Properties layer.

Your map should resemble the one below. When it does, you are ready to start working with Layout View.

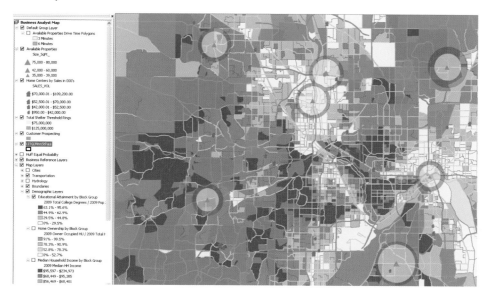

3.   Click View, then click Layout View (or click the Layout View button at the bottom left of the map) to open Layout View.

Business Analyst Desktop switches to Layout View, which should resemble this.

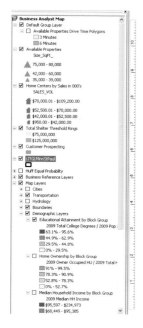

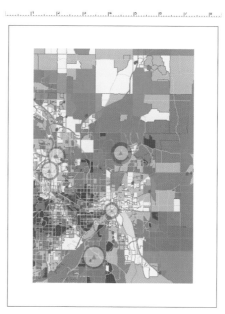

4. Click the Change Layout button at the far right of the Layout toolbar to open the Select Template window. Click the Business tab to view this set of templates. Select them one at a time to preview them. Select Analysis Landscape Letter.mxt, then click Finish to close the window and integrate map content into this template. It should resemble this.

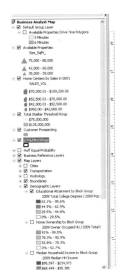 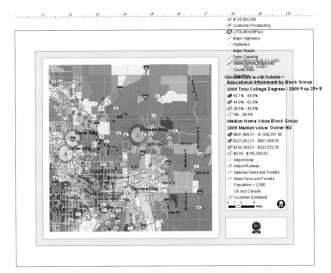

The map appears properly, but the legend is far too large. You will adjust it by editing its properties.

5. Click on the legend until a box with a dashed blue line boundary appears around it. Right-click, then click Properties to open the Legend Properties window. In the Legend tab, delete the default title **Legend**, then Click the Items tab. Click on the Items tab and click the double left arrow between the Map Layers box and the Legend Items box to move all items to the left and empty the Legend Items box. Click Apply to see the effect. Select the four layers you wish to include in the legend (Available Properties, Customer Prospecting, Total Shelter Expenditures Threshold Rings, and Educational Attainment by Block Group). Using the right arrow buttons between the boxes, move them to the Legend Items box. Click Apply to see the changes.

The Legend Properties box and map legend should resemble those on page 125. When they do, click the Legend button, delete Legend in the Title box, then click OK to apply the changes.

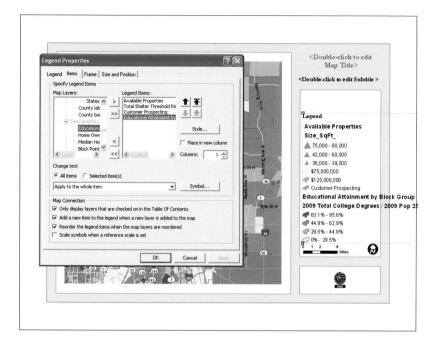

You may resize the legend to move into the allocated space. You may also move it to the top of the white box and resize the box to fit it, if you wish.

6. Double-click the text which reads **<Double-click to edit Map Title>**, then enter Living in the Green Lane **Property Options for First Store** in the Text box. Separate this title into two lines using the Enter key. Click OK.

7. Repeat this procedure with the Subtitle text, entering **With Sales Thresholds and Best Customer Prospects** as the text, also on two lines. Click OK.

    Your layout should resemble the one on page 126. When it does you are ready to export the map for inclusion in Living in the Green Lane's business plan.

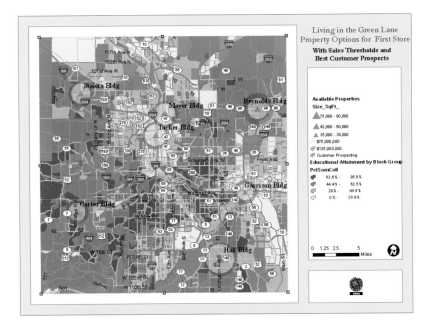

8. On the menu bar, click File, then Export Map to open the Export Map window. Select jpg as the file type, enter **LITGLAvailableProperties** as the filename, navigate to the folder C:\My Output Data\Projects\LITGL Minneapolis St Paul\CustomData\Chapter5\, and click Save. The map document is now ready to be inserted in LITGL's business plan with an Insert, Picture command sequence in Microsoft Word.

The second map will focus on the Steiers Building and its trade area. You will use the Median Home Value by Block Group thematic map as the background layer and display the Steiers Building, its 3- and 6-minute drive-time trade areas, nearby shopping centers and competing home centers above it.

9. While still in Layout View, zoom the map to the Steiers Building property. Toggle the layers in the Table of Contents to display the Available Properties, Available Properties Drive Time Trade Areas, Home Centers by Sales in 000's, Shopping Centers, and Median Home Value by Block Group layers. Adjust the transparency of the trade areas layer to improve the visibility of the Home Centers and Shopping Centers layers.

10. Edit the Properties of the Legend to include the five visible layers. Adjust its size and that of the box which surrounds it as necessary. Edit the Properties of the Title by entering **Recommended Site for Living in the Green Lane Store** as the new title. Edit the Properties of the Subtitle by entering **Steiers Building and Trade Area** as the new subtitle. Your Layout View should resemble the one on the facing page. When it does, click File, Export Map, and save the map in JPEG format as **RecommendedSite.jpg** in C:\My Output Data\Projects\LITGL Minneapolis St Paul\ CustomData\Chapter5\. It is now ready to be inserted into the Living in the Green Lane business plan.

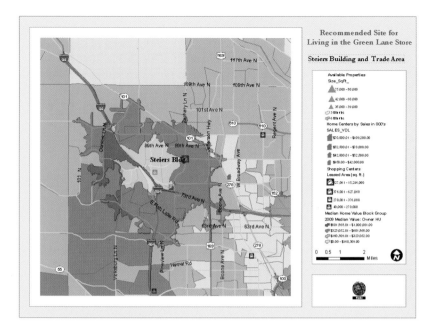

You have completed the site selection process, recommended the Steiers Building for the first store, and generated map documents to support that recommendation in Living in the Green Lane's business plan.

You have completed the task of selecting the site for Living in the Green Lane's first store. Business Analyst was an indispensable tool in that process, but did it increase the profitability of the decision? Let's reflect for a moment on the ROI associated with using Business Analyst for site selection in the absence of customer data.

ROI considerations

The return-on-investment equation is complicated in this case by the number of alternative sites under consideration. In the absence of the trade area and reporting functions of Business Analyst, it isn't certain which of the available properties would have been recommended. For that reason, the table below offers comparative figures for three sites. The Steiers Building is the recommended site. The Carter Building is the second of the two sites evaluated with detailed reports. The Garrison Building boasts the largest trade area in terms of population and households, and—absent the data on the green-consumer profile characteristics—might well have been the most attractive site.

Thus, this table seeks to present the financial benefits of selecting the Steiers site relative to the second most attractive site (Carter) and the site that might well have been chosen in the absence of customer profiling and lifestyle segmentation data (Garrison). It does so by estimating sales volume from green-profile customers and nontargeted customers separately.

Review the contents of the table. The first row shows the number of households in each trade area. The second row reports average household spending on LITGL's core product lines, remodeling, and major appliances. Note that additional services and new construction spending are not included; thus the table does not estimate total revenue.

The next three rows report the number of households in the trade area that are in the green-consumer profile, the estimated number of those households who will be LITGL customers, and the store's revenue from their purchases. The Tapestry Segmentation Report provides the data for classifying households as within the green-customer profile or outside it. This figure is based on a relatively modest 15 percent penetration rate in the first year for green-consumer households.

The next three rows make the same calculations for trade-area households that are outside the parameters of the green-consumer profile based on the Tapestry Segmentation Report. The market size and revenue rows assume a significantly lower 3 percent penetration rate for households outside the green-consumer profile.

The next row reports total projected revenue for each of the three trade areas. The final row reports decremental revenue for the Carter and Garrison trade areas. This is the value by which projected revenue in the Steiers trade area exceeds the figure for that trade area.

Attribute	Steiers site	Carter site	Garrison site
Households	13,631	12,363	73,114
Average spending per household	$5,319	$5,701	$2,174
Households in green profile	13,046	9,401	3,933
Assumed LITGL customers @ 15%	1,957	1,410	590
Revenue from households in profile	$10,409,283	$8,038,410	$1,282,660
Households *not* in green profile	315	2,962	69,181
Assumed LITGL customers @ 3%	10	89	2,076
Revenue from households *not* in profile	$53,190	$507,389	$4,513,224
Total revenue	$10,462,473	$8,545,799	$5,795,884
Decremental revenue		$1,916,674	$4,666,589

Table III.2 Revenue projections by trade area

The figures in the table illustrate the value of the reporting functions of Business Analyst Desktop. Remember that in initial screening, both the Carter and Steiers sites displayed significantly better values relative to the green-consumer profile than did Garrison. This poor match, low average home-related expenditures, and low purchasing indexes turned your attention away from the Garrison site despite the high number of households it contains. The wisdom of this decision is reflected in the significantly lower projected revenue from this site.

In the final selection process, the differences between the Steiers and Carter sites emerged from indications in the Market Profile Report that were illuminated more fully in other detailed reports as well as the Tapestry Segmentation Area Profile. Thus, the $1,916,674 increase in projected revenue generated by selecting the Steiers site instead of the Carter site is a direct benefit of these reports and your analysis. Similarly, the $4,666,589 increase in projected revenue relative to the more populous Garrison site is the direct result of the customer profiling and lifestyle segmentation analysis you performed.

These are the benefits. The incremental costs of this analysis are the acquisition costs for Business Analyst Desktop, the Segmentation Module extension, as well as your salary. While both these resources will serve Living in the Green Lane in a variety of future decisions, their combined costs are substantially exceeded by the difference in revenue between the two most attractive sites, and are dwarfed by the difference between the recommended Steiers site and the much more populous Garrison site.

With the Steiers site selected, your initial tasks are complete. As Living in the Green Lane grows and evolves, you will turn your attention to other applications of Business Analyst within the enterprise. Before doing so, however, review what you have learned in chapters 4 and 5.

Summary of learning

You have expanded your GIS knowledge by learning:

1. The value of integrating external data into GIS projects
2. The value and use of trade area generation and report creation
3. The types of trade area approaches that require no customer data or their relative value
4. How data on population characteristics, consumer expenditures, and lifestyle segmentation can be used to assess trade areas
5. The value of map documents as communication tools to support reports and business plans

You have enhanced your GIS skills by using Business Analyst to:

1. Geocode data on available properties and include it in a GIS project
2. Use customer prospecting tools and grid-based heat maps to identify concentrations of attractive customers
3. Create threshold trade area rings around locations based on consumer expenditure data
4. Create Huff Equal Probability Trade Areas and Locator Reports, and use them to assess the competitive environment of available locations

5. Create drive-time trade areas around available locations and create reports detailing their characteristics
6. Perform comparisons of site population, purchasing, and lifestyle segmentation characteristics by using general and detailed reports
7. Select and recommend a retail site by integrating and evaluating information from several maps and reports
8. Design map documents to illustrate your analysis and support your recommendation

Part IV

Customer profiling and site selection with customer data

Relevance	Successful businesses wishing to understand their customers' characteristics, values, and purchasing behaviors more fully and use this information to identify the most favorable sites for retail expansion.
Business scenario	LITGL must build on its success by expanding its product lines to meet the evolving expectations of green consumers and by identifying new retail locations to serve concentrations of green-lifestyle customers in its market area.
Analysis required	LITGL must geocode the customers in its Green Living loyalty club, identify the highest volume purchasers, and profile the demographic and lifestyle characteristics of this group of customers. The company then will use this customer profile to determine the most attractive additions to its line of green-living products and to select the best sites for two new green-lifestyle centers in the Minneapolis-St. Paul area.
Role of business GIS in analysis	Geocode Green Living members and identify high-volume purchasers. Use a spatial join procedure to assign demographic and lifestyle values to customers. Create demographic and lifestyle profile of high-value customers. Use Market Potential Indexes to determine customer values and buying habits, and to select appropriate new product lines to serve them. Use customer data to create sales-based trade areas around existing store. Create penetration and distance decay reports for existing store. Use Principal Components Analysis to rank opportunities presented by alternative available new store sites. Use Advanced Huff Analysis to project market penetration of selected site.
Integrated business GIS tool	ESRI Business Analyst Desktop and ESRI Segmentation Module extension.
ROI considerations: cost of business GIS	Business Analyst Desktop and Segmentation Module purchase, salary of GIS analyst.
ROI considerations: benefits of business GIS	Increased sales from well-selected new product lines. Increased sales from well-sited new green-lifestyle centers.
Environmental impact of business decision	Lower driving time and more efficient shopping for best customers. Broader line of green-lifestyle products in conveniently located retail centers.

Table IV.1 Executive summary

The Living in the Green Lane scenario

In its first two years, Living in the Green Lane's store has been a success.[1] It has exceeded projected sales, established a reliable customer base built around its Living Green loyalty club with almost 600 members, and established its position in the Minneapolis-St. Paul area as a champion of green building products and techniques.

As a result of this performance, Janice Brown and Steven Bent, cofounders of Living in the Green Lane, are in the enviable position of determining how best to build upon and extend this success. While crafting a growth strategy, they must weigh several relevant factors. Although sales have been solid, construction and home improvement are inherently cyclical businesses. In addition, many sales are project-oriented. This means that customers may have high volumes of purchases in years when they undertake projects and significantly lower levels of purchases in years when they do not. Additional fluctuations occur with economic cycles, when consumers adjust home-related expenditures to their income levels, as well as to governmental policies at various levels that provide tax incentives for environmentally beneficial renovations. For that reason, Janice and Steven wish to add product lines to LITGL's product mix that are less subject to such fluctuations.

Opportunities to do so arise from several directions. First, a need for maintenance and upgrade services often may follow significant renovation projects. Similarly, customers who have invested in green products and techniques in renovation projects often are attractive candidates for environmentally friendly lawn maintenance and/or pest-control services.

Second, for green consumers, green building is simply one element in a set of lifestyle factors that encompasses the broader concept of "responsible living." Other factors include a variety of wellness concerns such as the type, origin, and production methods of food products, fitness products and services, and personal-care products. Another stream of factors relates to nurturing and protecting local resources. This might include preferences for locally produced products, farmers' markets for food shopping, and support for local merchants, restaurants, and services. When looking beyond locally produced products, the "responsible living" concept embraces organic production techniques, fair-trade products, and products produced by companies committed to managing their own sustainability and environmental impact.

Third, green consumers also tend to embrace the concept of "low-impact living," which involves making conscious lifestyle changes to lower their environmental impact, particularly the carbon impact, of daily activities. Included here are daily transportation and home energy consumption decisions as well as travel decisions in favor of local vacation destinations, ecotourism, or humanitarian service projects as vacation options.

Fourth, for many green consumers, these value streams are not discrete but integrated, so that commitment to one often entails commitment to several others. Thus, these consumers tend to think of green living as something that extends beyond home building and renovation to a broad range of lifestyle patterns.

Janice and Steven believe that this complex set of values and behaviors offers the best growth opportunity for Living in the Green Lane. They believe that the concept of a green-home center is significantly less powerful for these customers than that of a green-lifestyle center. If Living in the Green Lane can find the right mix of products, services, and shopping environment to evolve into a green-lifestyle center, it can serve these customers more effectively and profitably. They further believe that integrated business GIS is indispensable to this effort and, for that reason, have enlisted you to help achieve this goal.

This is the task upon which chapters 6 and 7 focus. To achieve it you will identify LITGL's best customers, use a spatial overlay procedure to explore their demographic and lifestyle characteristics, and create a customer profile. In chapter 6 you will use this customer profile to identify the best merchandising opportunities for product line expansion and the best media for reaching target customers. In chapter 7 you will use the customer profile to create customer-based trade areas around LITGL's store and find sites in the Minneapolis-St. Paul area that match these profile characteristics. Using these techniques, you will select sites for two new stores.

Integrated business GIS tools in customer profiling and site selection with customer data

The customer profiling process in chapter 6 begins with a Microsoft Excel table listing the addresses of members of the Living Green loyalty club and their purchases from LITGL in the past calendar year. The first business GIS task will be to geocode these addresses. Using the ESRI Business Analyst Desktop locator service, you will assign each customer address latitude and longitude values, then use these values to locate the customer on a map. This service uses point, street, and ZIP Code locator references to obtain the most accurate location estimate possible for each address, and adds a field to the resulting attribute table to indicate the precision level at which each customer address was matched.

The next step is to identify the most attractive segment of customers—those with the highest level of purchases in the past year. This approach is based on the general principle that a relatively small percentage of a company's customers account for a relatively larger percentage of its sales. When customer sales follow a Pareto curve, 20 percent of customers produce 80 percent of sales. In the product usage approach to segmentation, this top 20 percent would be the high users' segment [2]. Another approach is to divide users in half at the median value of purchases and define these as heavy- and light-using segments. Database marketing systems perform analyses based on the recency, frequency, and monetary values of purchases to assign customers to cells based on their overall value.[3] Similarly, customer relationship management systems use estimates of lifetime purchases of customers minus acquisition and retention costs to estimate lifetime customer value. Customers are then segmented based on their value to the company.[4]

Business Analyst Desktop provides several tools for segmentation based on purchasing levels. Symbology and classification tools allow users to create quantile classifications of customers or determine natural breaks in the distribution of sales. Statistical tools provide both quantitative and visual representations of sales distribution. Finally, customer-derived trade areas create polygons encompassing designated percentages of customers or customer purchases. You will utilize several of these tools to determine the composition of the High Purchases segment that will be the basis of your customer profile.

When high-volume customers have been identified with the most appropriate approach, the next step is to assign demographic and lifestyle values to each customer based on his or her geographic location. You may perform this function using either a spatial join procedure or the Spatial Overlay Analysis tool in Business Analyst Desktop. You will use the former approach, which extends the functionality of the shapefile you created in chapter 3, to assign each customer record the values of the block group in which it is located. Clearly, these values are estimates of general patterns across customers rather than exact values for individual customers. As such, they are quite useful in capturing the characteristics that distinguish between large groups of customers.

While these tools are straightforward in implementation, care must be taken in their use. Some attributes such as median age or average household size make sense when assigned to individual records. Others, such as total college graduates or total home owners, do not. In these cases, using attributes that express these measures as a percentage of the population or as a percentage of households is more useful. The percentages are best understood as probabilities that the person is a college graduate or owns a home. When repeated for a large number of customers in a segment, the resulting values are indicative of general patterns in the segment.

For Tapestry Segmentation segment definition, each customer is assigned the dominant Tapestry segment for the relevant geographic unit. In Business Analyst Desktop, census tracts are the smallest units for which Dominant Tapestry Segmentation segments are assigned. However, with the Segmentation Module installed as an extension to Business Analyst Desktop, Dominant Tapestry Segmentation segments are recorded at the block group level, the value at which they are actually assigned. In this situation, the Tapestry Segmentation segment assignment for each household is very precise.

Once demographic and lifestyle values are assigned, you will use the Business Analyst Desktop Summarize function to calculate a summary table that reports the values of these attributes by market segment. When applied to demographic attributes, this process produces a summary table that allows users to determine the distinctive characteristics of each segment. When applied to the Dominant Tapestry Segmentation segment attribute, it allows users to determine the concentration of customer groups by Tapestry Segmentation membership. In each case, comparison between segments, and between segments and national and/or regional characteristics, allows users to create customer segment profiles of the distinguishing characteristics of attractive segments.

While this process produces significant insight into customer characteristics, its power is multiplied exponentially when integrated with the Market Potential Indexes based on consumer survey data compiled by Mediamark Research and Intelligence LLC. These indexes are reported for all 65 Tapestry Segmentation segments across a broad range of questions that inquire about the values, activities, preferences, and purchasing behavior of consumers in the United States. For a particular behavior, say purchasing lawn-maintenance services, the index for each Tapestry Segmentation segment reports its behavior relative to the national average. Thus an index value of 80 means that this segment is 20 percent *less* likely than average to purchase lawn-maintenance services while an index of 120 means it is 20 percent *more* likely to do so.

You will use Market Potential Indexes to study the purchasing preferences of the Tapestry Segmentation segments most prevalent in LITGL's High Purchases segment. You will then design product line, merchandising, and communication strategies to serve these customers more effectively.

Implementation of these strategies will enable Living in the Green Lane to reposition itself from a green-home center to a green-lifestyle center. This will expand profitability through increased sales to green customers in LITGL's trade area. It will also provide the concept and customer profile that you will use to identify the best site for the second and third Living in the Green Lane stores in the Minneapolis-St. Paul area.

That store selection is the focus of chapter 7, which covers the site selection process using customer data.

Trade area options with customer data

Beyond the profiling process, customer data also enables refinement of trade area analysis and site selection procedures. To this point, you have created trade areas using geographic, population, and competitive factors. With the addition of geocoded customer records, you may extend these efforts by integrating sales information from Living in the Green Lane's best customers.

You will begin by creating trade areas based on the geographic distribution of customer data. Specifically, you will create polygons that encompass specified levels of sales data. Business Analyst Desktop provides the option to base these areas on the percentage of customers they contain or some other weighted attribute such as number of orders or sales volume. As you are interested in the concentration of sales, you will use sales volume for this purpose and specify polygons that contain 50 percent and 80 percent of sales volume, respectively. Comparison of these trade areas to the drive-time and expenditure threshold polygons you designed in chapter 5 will allow you to evaluate those projections of retail attractiveness to customers in the area.

With these trade areas in place, you will use Business Analyst Desktop to evaluate them for market penetration and distance decay. Market penetration measures LITGL's customers as a percentage of households in nearby geographic areas. High levels of penetration indicate market areas nearing saturation through effective marketing efforts. Low levels of penetration indicate market areas that are ripe for expanded sales with appropriate marketing strategies. The results of this analysis is a map that depicts penetration levels and a report listing precise penetration rates by market area.

Similarly, Business Analyst Desktop enables you to measure distance decay within the market area. Recall the assumption in the basic Huff model in chapter 4 that the attractiveness of a retail location decreases as the distance from the location increases. To assess this effect, you will create drive-time trade areas similar to those in chapter 5 and measure the number of customers in each of the resulting polygon trade areas. You will also augment this map with an accompanying report. This will help you estimate the impact of this measure more precisely as you select sites for Living in the Green Lane's second and third stores.

With the insights developed in the customer profiling and customer-based trade area analyses, you are ready to seek out the most attractive new retail sites for LITLG's new market model. To do so you will use the Find Similar tool and Principal Components Analysis to rank available sites based on their similarity to a defined customer profile. In this process, you will set parameters for the attributes and criteria you wish to include in the analysis. Business Analyst Desktop will use those specifications and the multivariate Principal Components Analysis technique to rank the available sites on their attractiveness relative to these criteria.

Having ranked the available sites relative to your desired customer profile, you will use the Advanced Huff Analysis procedure to estimate sales of the most highly ranked sites relative to competing stores. To do so, you will integrate data on desirable store characteristics in addition to distance data to project sales in the market area at the block group level. This projected sales data provides the information necessary to select the sites for two additional Living in the Green Lane stores.

Chapter 6

Building a profile of distinctive customer characteristics

Janice and Steven's first growth strategy is to learn more about Living in the Green Lane's customers and use that information to serve customers more effectively and expand LITGL's product mix with attractive new product lines. You will use Business Analyst Desktop tools to complete that demographic and lifestyle profile and identify these new marketing opportunities.

Run Business Analyst Desktop; geocode customer list

1. Click Start, Programs, ArcGIS, Business Analyst, BusinessAnalyst.mxd to run ArcMap, load the Business Analyst Extension, and then load the default Business Analyst map.

2. Click OK when the Update Spatial Reference dialog box appears, then close the Business Analyst Assistant window on the right of the screen.

3. Click File, then click Open. Navigate to C:\My Output Data\Projects\LITGL Minneapolis St Paul\CustomData\ChapterFiles\Chapter6\LITGLCustomers.mxd. Click the map file to open it.

    Your screen will resemble this below. The Table of Contents includes a layer for the Living in the Green Lane store as well as home centers and shopping malls. There is also a thematic layer depicting Median Home Value at the block group level. This layer is based on a customized shapefile similar to the one you created in chapter 3.

4. Click the drop-down arrow on the Business Analyst toolbar, then click Customer Setup to initiate the Customer Setup wizard. In the first window, select Create New Customer layer, then click Next. In the resulting window, select Tabular data, then click Next. In the resulting window, select In a file on my computer, then click Next. Click the Open file button and navigate to C:\My Output Data\Projects\LITGL Minneapolis St Paul\

CustomData\ChapterFiles\Chapter6\LITGLCustomers.xls. Double-click this file to view its contents, then select the Customers table and click Add. Click Next.

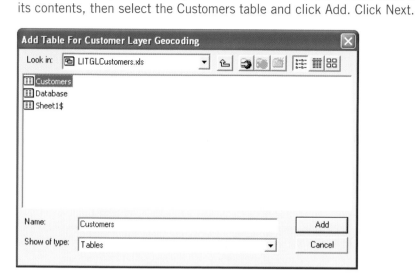

5. Click Next to view the Input Fields window. Review the settings and accept the default values in the Address, City, State, and ZIP fields and click Next. In the resulting window, confirm that Cust ID is selected in the Name field and "none" is the setting in the StoreID field. Click Next. In the final window, enter **LITGL Customers** in the name field. Click Finish.

Business Analyst Desktop geocodes the customer addresses in the table, displays them on the map, and adds a layer to the Table of Contents, which uses a single symbol to depict all customers.

Right-click the LITGL Customers layer, then click Open Attribute Table to view the customer records. From the left, all the attribute fields in each record were assigned by the geocoding process. They report the level and precision of the geocoding operation as well as the exact address matched by the geocoding service. Cust ID is the first field from LITGL's loyalty club customer records.

Move to the far right of the table to the field LYPurchase, which reports total dollar purchases of each customer at Living in the Green Lane last year. You wish to identify those customers with the highest levels of purchases and use them to develop the customer profile. You will calculate summary statistics of this field to obtain sales data.

Analyze purchasing data to define high purchases segment

1. Right-click the LYPurchase label at the top of this field, then click Statistics to calculate Summary statistics.

The Statistics window should resemble this one. The Sum value is total purchases by loyalty club customers while the Mean value reports average purchases per customer. Note the wide variation between the Minimum and Maximum values, a range reflected in the high Standard Deviation as well as the Frequency Distribution chart, which displays a bimodal distribution, indicating significant variations in purchasing patterns. The cluster of customers in the higher spending group appears to begin at roughly the $16,400 level of annual purchases. You will use this figure as the threshold level for defining the High Purchases segment.

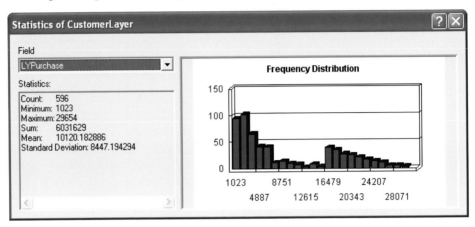

2. Close the Statistics window. Click Selection on the main menu bar, then Select by Attributes to open the Select by Attributes window. Scroll down to the bottom of the attribute list and double-click LYPurchase to add this field to the Expression box. In that box enter **> 16400**. The Select by Attributes window should resemble the one below. When it does, click Apply to select the appropriate customer records. Click Close to close the window.

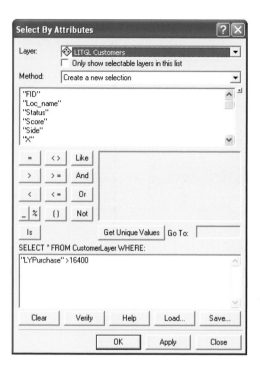

The customer features with the highest level of annual purchases are selected. You will use the Add Field and Field Calculator functions to assign them to the High Purchases segment.

3. If the LITGL Customers layer attribute table is not visible, expand it. Click the Options button at the bottom of the table, then click Add Field to open the Add Field window. Enter **PurchSeg** as the name of the field, select Text in the Type box, and enter **15** as the Length. The window should resemble this. When it does, click OK to add the field to the far right of the table.

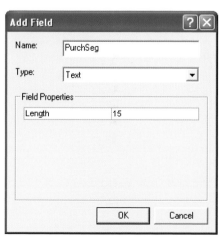

Move to the right side of the attribute table where you will find the new field with its label at the top of a blank column. You will use the Field Calculator function to assign the selected customers to the High Purchases segment.

4. Right-click the PurchSeg label, then click Field Calculator. When the Calculate Outside Edit Session warning box appears, click Yes. In the box just below PurchSeg =, enter "**High Purchases**". Be sure to include the double quotation marks to indicate that you are entering text. The window should resemble the one below. When it does, click OK.

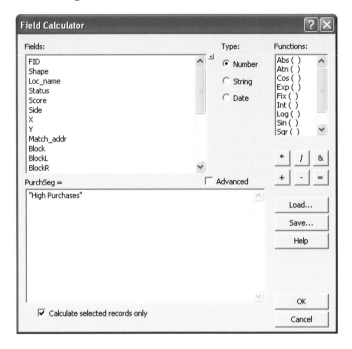

The selected customer features are assigned to the High Purchases segment as indicated in the PurchSeg field. You will assign the remaining customer features to the Low Purchases segment in the same way.

5. Click Options at the bottom of the attribute table, then click Switch Selection to reverse the selection in the table and to select customer features with annual purchases below the defining threshold.

6. Right-click the PurchSeg label, click Field Calculator, enter **"Low Purchases"** complete with double quotation marks in the expression box, and click OK. Click Options, Clear Selection to remove the selection. Minimize the attribute table.

The selected customer features are assigned to the Low Purchases segment. All features have now been assigned to a segment. You will display their segment classification on the map of Living in the Green Lane's store.

7.   Open the Layer Properties box for the LITGL Customers layer. In the Symbology tab, select the Categories: Unique values option in the Show: box. Select PurchSeg as the Value Field. Click the Add All Values to add both segments to the map. Unselect the *all other values* option. Choose a color ramp of your choice to distinguish customers in the two segments by the color of their dot on the map. The Symbology box should resemble this. When it does, click OK to close the Layer Properties box and apply the changes.

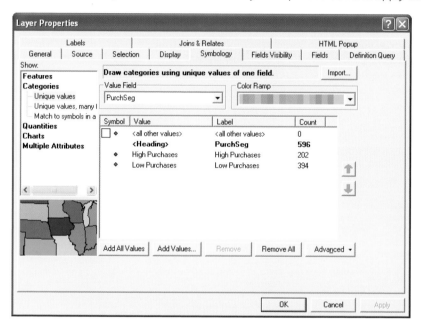

All customers are now displayed on the map in colors that correspond to the purchasing segment to which they have been assigned. This map allows you to explore the geographic distribution of these purchasing segments relative to Living in the Green Lane's store.

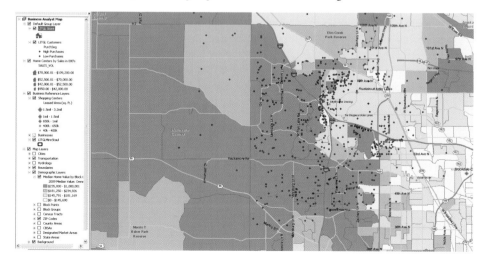

You now wish to compare the demographic and lifestyle characteristics of the segments in order to identify the distinguishing features of the High Purchases segment profile. You will perform an overlay operation using the spatial join function to do so.

Compare segment characteristics with spatial join and summarize procedures

A Spatial Join is a procedure that appends data from one layer to another based on the spatial relationship between features in the layer.[5] In this instance, you wish to understand the demographic characteristics of LITGL's customers more thoroughly. The demographic information you need is in the Median Home Value by Block Group layer. You wish to attach this data to features in the LITGL Customers layer. Specifically, each customer record will receive the demographic and Tapestry Segmentation values for the block group in which it is located. These values represent estimates of demographic and lifestyle characteristics based on the location of each customer.

1.   Right-click the LITGL Customers layer, click Join and Relates, then Join to open the Join Data window.

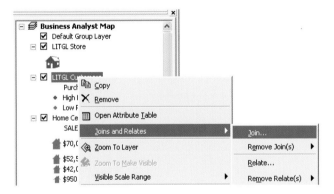

2.   In the drop-down box, select Join data from another layer based on spatial location. In the Choose layer field use the drop-down box to select Median Home Value by Block Group as the layer to join to the customer layer. In the file locator field at the bottom of the box, navigate to C:\My Output Data\Projects\LITGL Minneapolis St Paul\ CustomData\ChapterFiles\Chapter6\ and designate **LITGLCustomerDemo.shp** as the name of the new customer layer. When the window resembles the one on the next page, click OK.

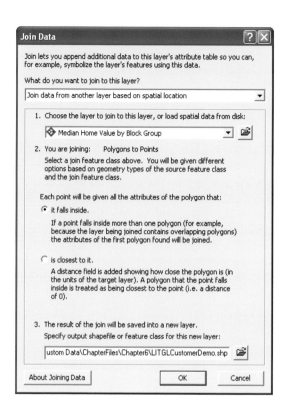

Business Analyst Desktop appends the appropriate demographic and Tapestry Segmentation data to each customer record and creates a new layer, LITGLCustomerDemo, which includes both customer and demographic attributes.

3. Use the Symbology tab in the Layer Properties window for the LITGLCustomerDemo layer to match the symbology of the LITGL Customers layer. When they are identical, right-click the LITGL Customers layer, then click Remove to remove it from the Table of Contents and map.

    The LITGLCustomerDemo layer is now ready for use in creating summary tables of customer characteristics.

Create summary table of segment characteristics

You wish to compare the characteristics of the two segments of LITGL customers. You will do so with the Summary function of Business Analyst Desktop, which allows you to calculate summary tables that display those characteristics aggregated for the customers in each segment.

1. Right-click the LITGLCustomerDemo layer and click Open Attribute Table. In the table, right-click the PurchSeg label at the head of the column for that field. Click Summarize to open the Summarize dialog box.

2.   Confirm that PurchSeg is selected as the field to summarize. In the next box, expand the attribute LYPurchase (Purchases from LITGL in past year) and select the sum option. Select the following fields and aggregation options in the middle box.

AVGHHSZ_CY   - Average   Average household size in current year
LYPurchase   - Sum   Purchases from LITGL in past year
MEDHINC_CY   - Average   Median household income in current year
MEDVAL_CY   - Average   Median home value in current year
CYPctOwnHm   - Average   Percent home ownership in current year
FYPctOwnHm   - Average   Projected percentage of home ownership in five years
PctCollDeg   - Average   Percent adults with college degree in current year

In the Specify Output Table box, navigate to C:\My Output Data\Projects LITGL Minneapolis St Paul\CustomData\ChapterFiles\Chapter6\, enter **PurchaseSegmentSummary.dbf** as the filename and dBASE table as the file type. Click Save. The Summarize dialog box should resemble the one below. When it does, click OK. When given the option to add the result table to the map, click Yes.

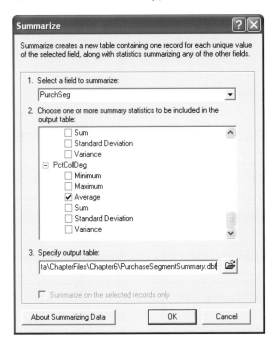

Business Analyst Desktop calculates summary values for the two segments and adds the results to the map as a data table below the LITGL Customer Demo layer. (Note: You will not be able to see the table unless the Sources tab at the bottom of the Table of Contents is selected. If it is not, click that tab and the table and its path will appear under the Customer Overlay layer.)

Open the data table from the Table of Contents in the Source mode and compare the values for the High Purchases and Low Purchases segments. Use these values to complete the following table and answer the question below it. Notice that you must calculate the percentages of customers and purchases in the second and fourth rows.

Population characteristic	High purchases	Low purchases
Customers in segment		
Percent of total customers (calculate)		
Segment total purchases		
Segment purchases as percent of total purchases (calculate)		
Average household size in current year		
Median household income in current year		
Median home value in current year		
Percent home ownership in current year		
Projected percentage of home ownership in 5 years		
Percent adults with college degree		

Table 6.1 Population characteristics of high purchases and low purchases segments

How do the High Purchases and Low Purchases differ from each other? How are these differences related to the characteristics of green-consumer segments?

Create summary table of Tapestry Segmentation segments

To understand the High Purchases segment more fully, you wish to determine the Tapestry Segmentation segments that form it. You will use the Summarize function on the customers in this segment to determine its lifestyle segment composition.

1. Right-click the LITGLCustomerDemo layer and click Open Attribute Table. Click Options. Use the Select by Attributes function to select those customers in the High Purchases segment by entering the following query statement in the lower box: **"PurchSeg" = 'High Purchases'** (note the double quotes on the left and the single quotes on the right). Click Apply.

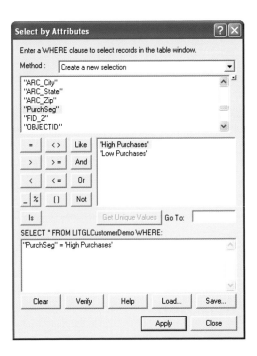

2. In the LITGLCustomerDemo attribute table, right-click the Dominant Tapestry Code label at the head of the column for that field. Click Summarize to open the Summarize dialog box. Confirm that DOMTAP is selected as the field to summarize. In the next box, expand the attribute LYPurchase (Purchases from LITGL in past year) and select the Sum and Average aggregation options.

3. Click the folder icon to the right of the Specify Output Table box to open the Saving Data window. Navigate to C:\My Output Data\Projects\LITGL Minneapolis St Paul\ CustomData\ChapterFiles\Chapter6\, enter **HighSegTapestrySummary.dbf** as the filename and dBASE table as the file type. Click Save to return to the Summarize window. Confirm that the Summarize on selected records only option is selected. (Note: If this option is not enabled, confirm that you performed step 1 above correctly and that the customers in the High Purchases segment are selected.) The Summarize dialog box should resemble the one on page 151. When it does, click OK. When given the option to add the result table to the map, click Yes.

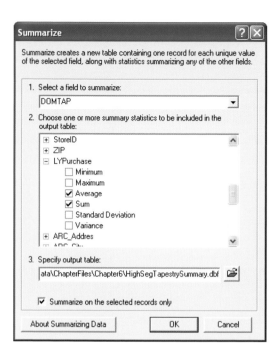

Business Analyst Desktop calculates the values for the Tapestry Segmentation segments in this purchasing category and adds the resulting table to the Table of Contents below the LITGLCustomerDemo layer in the Source Mode. Open the data table and compare the number of customers and purchasing patterns of the Tapestry Segmentation segments. Use these values to complete the table on page 152 and answer the questions below it.

4. To preserve your work, save your map file as **LITGLCustomers2.mxd** to C:\My Output Data\Projects\LITGL Minneapolis St Paul\CustomData\ChapterFiles\Chapter6\.

Tapestry Segmentation Neighborhood Segment	# of High Purchasers	% of High Purchasers	% of U.S. population[6]	Average purchases	Total purchases
04: Boomburbs			6.7%		
06: Sophisticated Squires			8.2%		
02: Suburban Splendor			5.2%		
12: Up-and-Coming Families			9.7%		
13: In Style			6.7%		
07: Exurbanites			6.1%		

Table 6.2 Tapestry Segmentation composition of high purchases segment

Which Tapestry Segmentation segments are the most numerous in LITGL's High Purchases market segment?

How do their concentrations in this segment compare to national averages?

Which segments have the highest level of average household purchases in the past year?

Which segments have the highest level of total purchases from LITGL in the past year?

Profiles of the dominant Tapestry Segmentation segments among LITGL's best customers provide significant insight into their characteristics, values, and purchase preferences. The Market Potential Indexes from Mediamark provide even greater insight in its annual statistics on the values, activities, and purchasing behavior of Tapestry Segmentation segments.

Market Potential Indexes report the responses of each segment to hundreds of questions about their market behavior. The results are reported as indexes based on a value of 100. An index of 100 for a segment relative to a specific behavior means that this particular segment reports the behavior at exactly the same rate as the national average. A value of 125 means that the segment's rate is 25 percent above the national average, while a value of 75 means that it is 25 percent below the national average.

Appendix A is a table containing a series of behaviors selected for their relevance to the green-lifestyle customer Janice and Steven wish to reach as well as MPI values for the six most numerous Tapestry Segmentation groups in High Purchase segment. Review the values for LITGL's major Tapestry Segmentation segments and use them to answer the questions on page 152.

Based on the values in appendix A, how well do these Tapestry Segmentation segments fit the following values, behaviors, and media consumption patterns of the green-lifestyle consumers Janice and Steven wish to target?

Purchase lawn and garden maintenance services? (See Lawn and Garden in appendix A)

Purchase pest control services? (See Lawn and Garden)

Concern for environmental issues? (See Civic Activities, Lawn and Garden)

Interest in physical activities and fitness and wellness products? (See Apparel, Health, Leisure Activities and Lifestyle, Sports, and Travel)

Interest in fresh, organically produced fruits and vegetables? (See Grocery, Health)

Read magazines on environmental, health, wellness, and home-related topics (See Media)

Responsive to direct marketing and the Internet, including purchasing products online, by mail order, and phone? (See Internet, Mail and Phone Orders/Yellow Pages)

Which media are most appropriate for reaching these segments? (See Media)

Marketing strategy decisions

With this customer profile and your recommendations in mind, Janice and Steven have chosen several new products and services to add to the Living in the Green Lane mix. They believe that these additions will transform LITGL into the green-lifestyle center these customers are seeking.

Product or service	Projected revenue impact
Organic landscaping, lawn, and garden care service	Direct revenue from initial treatment and service
Organic pest-control service	Direct revenue from installation and service
Low-energy-consumption products department	Direct sales revenue
Organic food and personal-care products department	Direct sales revenue
Lease space to organic coffee shop and café	Lease revenue, greater customer traffic
Add wellness, health, and lifestyle titles to current department of home-improvement magazines	Direct sales revenue
Use part of parking lot for local farmers' market	Some direct revenue from stall rental, greater customer traffic
Use mulch from recycled materials to build walking track around parking lot	No direct revenue, greater customer traffic
Make product demo and seminar room available for local groups and/or companies for: • Yoga or aerobics sessions • Low-impact living seminars • Ecotourism seminars • Health and wellness seminars • Community development seminars	No direct revenue, greater customer traffic

Table 6.3 Product/service additions and projected revenue impact

With these additions, the Living in the Green Lane green-lifestyle center is ready for expansion to new stores. Since this concept and product mix are specifically crafted to appeal to a narrowly focused customer group, it is vital that new stores be located near concentrations of people with similar characteristics. For this reason, the customer profile you created must be a central component of the site selection process for new stores. This is the task upon which you will focus in chapter 7.

Chapter 7

Customer-based trade-area analysis and site selection

Janice and Steven's first growth strategy is to learn more about Living in the Green Lane's customers and use that information to serve customers more effectively and expand LITGL's product mix with attractive new product lines. You completed those tasks in chapter 6.

Their second growth strategy is to open new stores that deliver this enhanced product/service mix to areas whose population matches the High Purchase customer profile. In this chapter you will use the trade-area functions and Find Similar tools of Business Analyst Desktop to identify attractive sites for new LITGL stores.

Run Business Analyst Desktop; load map

1.   Click Start, Programs, ArcGIS, Business Analyst, BusinessAnalyst.mxd to run ArcMap, load the Business Analyst Extension, and then load the default Business Analyst map.

2.   Click OK when the Update Spatial Reference dialog box appears, then close the Business Analyst Assistant window on the right of the screen.

3.   Click File, click Open. Navigate to C:\My Output Data\Projects\LITGL Minneapolis St Paul\Custom Data\ChapterFiles\Chapter7\LITGLExpansion.mxd. Click the map file to open it.

This is similar to the map you designed at the end of chapter 6. It contains a layer for the LITGL store, a layer displaying available expansion sites (which is turned off), a LITGL Customers layer displaying Green Living Club members, a layer displaying competing home centers, a drive-time trade-area layer (which is turned off), and a thematic layer depicting Median Home Value based on the customized shapefile you created in chapter 3. There is also a study area layer, Mason site, that you will turn on and use later in this chapter. You will use these resources to continue your analysis and select a site for the second and third LITGL stores.

Create customer-derived trade area for Living in the Green Lane store and generate reports

1.   Click the drop-down arrow in the Business Analyst toolbar, click Trade Area. Select Create New Trade Area, click Next. Select Customer Data Required, click Next. Select Customer Derived Areas, click Next. Select LITGLStore as the store layer and ID as the ID field. Select the All Stores option, click Next.

2. In the resulting window, select LITGLCustomers as the Customer layer and StoreID as the Store assignment field, then click Next. In the resulting window, select the By a weighted value option and designate LYPurchase as the weighted field. The window should resemble this:

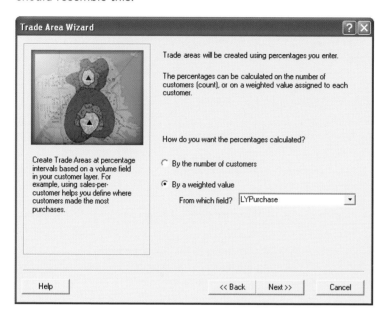

These are important settings in designing customer-derived trade areas. The first option, By the number of customers, would create polygons encompassing designated percentages of total customers. In contrast, the second option, By a weighted value, employs a user-designated attribute to seek concentrations based on some other value. In this case, that value is sales, which means that the trade areas you create will encompass designated percentages of total sales, thus assigning more weight to customers with high purchase levels than to those with lower levels. This is consistent with the customer profiling task you just performed, which concentrated on profiling customers with high levels of purchases.

3. When your window matches the image above, click Next. In the resulting window, select the 2 as the number of trade areas to create and enter **50** and **80** as the percentage of sales to include in the inner and outer trade-area polygons, respectively. The window should resemble the one on page 158. When it does, click Next.

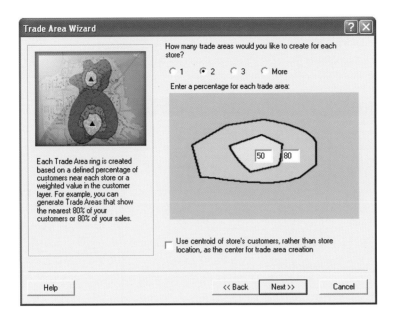

4. In the resulting window, select each of the available options and read its description below the sample map. Select the Detailed With Smoothing option, then click Next. In the resulting window, enter **Percent of Sales Trade Areas** as the name of the trade areas, select the Create Reports option, then click Next. In the resulting window, select the For Individual Features option, then click Next.

5. The resulting Report Templates window lists the available report templates in the box on the left. To select a report, simply move it from the box on the left to the box on the right with the arrow keys between the two boxes. Click the Market Profile and Tapestry Segmentation Area Profile reports to select them, then click the right arrow between the boxes to move them to the box on the right and order these reports. Click Next.

6. In the final window, select only the View Reports on Screen option. (You can export a report to a data file in several formats after your initial reading.) Click Finish.

Business Analyst Desktop creates the trade areas, adds them to the map as a layer and produces the reports you ordered. The reports will open in the center of your screen, obscuring the map layer. Compare the characteristics of the High Purchasing segment you profiled in chapter 6 with those of the Percent of Sales Trade Areas. If the demographic characteristics and Tapestry Segmentation composition are similar, there are likely significant opportunities to enhance sales by increasing market penetration within the trade areas. If these characteristics differ, the stronger growth opportunity is to seek market areas with higher concentrations of households with characteristics that match the customer profile.

You will use Trade Area Penetration functionality that follows to determine the exact levels of customer penetration you have achieved in the Percent of Sales Trade Area.

Close the report to view the trade areas displayed on the map, which should resemble the one below. Toggle the display of the Percent of Sales trade areas and the Drive Time trade areas you created in chapter 5. If the drive-time trade areas are the larger of the two, you have overestimated the attractiveness of LITGL's store. If the Percent of Sales trade areas are larger, you have underestimated the attractiveness. That is, in the latter situation you are drawing customers from a larger geographic area than you anticipated, indicating a more attractive retail concept. This, in turn, means that you can use a more extensive trade-area definition to search for sites for the second and subsequent stores.

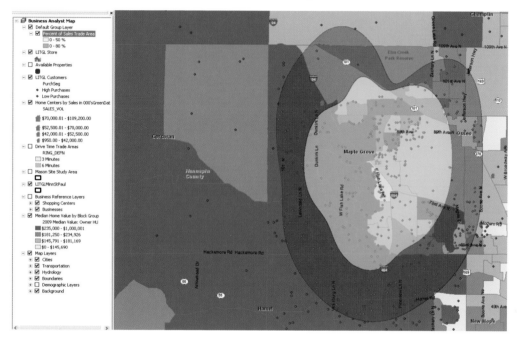

You will use Distance Decay functionality below to explore the relationship between distance from store and sales levels.

Create trade-area penetration map and report

1. Turn off the Drive Time Trade Areas layer. Click the drop-down arrow in the Business Analyst toolbar, then click Trade Area. Select Create New Trade Area, then click Next. Select Customer Data Required and click Next. Select Trade Area Penetration, then click Next. Select LITGLCustomers as the customer layer, and confirm that the Use a weight field in the customer layer option is NOT selected. Select StoreID in the Select Store ID field in Customer Layer. Click Next.

2.   In the resulting window, select Percent of Sales Trade Areas as the trade-area layer. Select Area_ID and Area_Desc as the trade-area ID and Name fields, respectively. When the window resembles the one below, click Next.

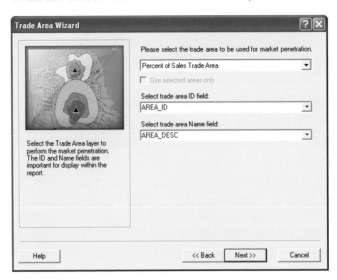

3.   In the resulting window, select StoreID in the Select Store ID field in Customer layer box and the ID field in the Select ID field in Trade Area layer box. Click Next.

4.   In the resulting window, select the Calculate using Business Analyst Data option, then select Standard Business Analyst Data (Generalized) as the data layer and Current Year Total Households as the data field. When your screen resembles the one below, click Next.

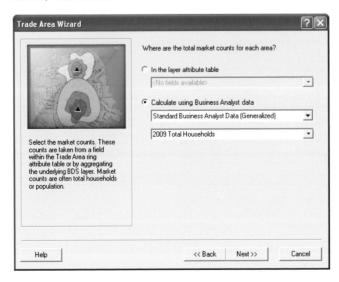

You have specified, with these settings, the customer and trade-area layers you wish to study. In addition, you have specified that you wish to view penetration values for LITGL's customers as a percentage of households in the trade area. This will indicate how fully you have penetrated the potential customer base of the market area.

5. In the resulting window, select the Generate Report and View report options. Enter **Trade Area Penetration Report** as the Report name and **Trade Area Penetration** as the Trade Area Name. Click Finish.

    Business Analyst Desktop determines the trade areas and displays them on the map. This layer is identical to the Percent of Sales Trade Area in Shape, but displays the penetration values of each polygon in the legend. In this case penetration is defined as the number of LITGL customers divided by total households in the trade area.

    Business Analyst Desktop also displays the Trade Area Penetration Report, which reports penetration rates in a table as well as a graph. These penetration rates are based on the number of customer households relative to total households. You also wish to compare penetration rates measured in terms of purchases, so you will revise the Trade Area Penetration wizard to include slightly different settings.

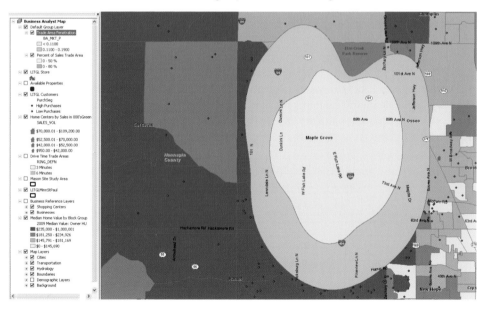

6. Click the drop-down arrow in the Business Analyst toolbar, then click Trade Area. Select Manage Existing Trade Area(s), and click Next. In the resulting window, select Trade Area Penetration, then select the Modify Trade Area option. Select Customer Data Required. Click Next. Choose Trade Area Penetration. Click Next to revise the Trade Penetration Trade Area Wizard by repeating the steps you performed earlier but with the new settings described on the next page.

7. In the Customer layer designation window, retain LITGL Customers as the customer layer. Select the Use a weight in the customer layer option and designate LYPurchase as the weighted field and StoreID in Select Store ID field in Customer layer. Click Next.

8. In the next window, select Percent of Sales Trade Areas, AREA_ID and AREA_DESC as the settings, then click Next. In the next window retain the StoreID and ID settings, and click Next. In the next window, retain Standard Business Analyst Data (Generalized) as the data source and select Shelter: Tot as the data field. Click Next.

9. In the next step, enter **Trade Area Penetration by Sales Report** as the Report name and **Trade Area Penetration by Sales** as the Trade Area Name. Click Finish to run the analysis.

With these new settings, you have specified that you wish to view penetration values for a weighted field, LYPurchase (purchases in the past year) as a percentage of total consumer expenditures on shelter. This is consistent with the pattern of concentration on sales levels from the profiling task as well as the Percent of Sales Trade Area approach. Business Analyst Desktop revises the map layer to reflect these settings and calculates a new report.

Review the report, which resembles the one below. Like the initial one, it displays the number of customers and percentage of total customers in each of the defined trade areas. It also displays total purchases and percentage of total purchases for each of the trade areas. In this report, however, the penetration rate is calculated using total customer purchases as a percentage of total shelter expenditures in the trade areas. High numbers indicate a strong sales presence by LITGL, while lower numbers indicate that greater market penetration is possible if the profile of the market area matches that of high-volume customers. Recall that Shelter: Tot is the broadest category of housing-related expenditures and includes categories LITGL does not offer. As a result, the company's market penetration will be understated.

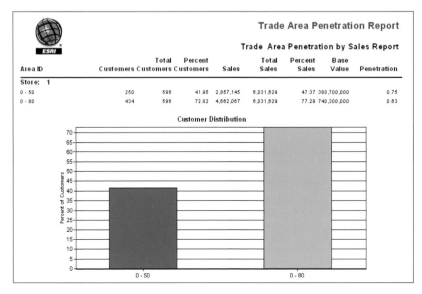

Create distance decay map and report

Distance Decay trade areas report penetration rates based on numbers of customers and/or customer purchases for trade areas defined by distance. Thus, they are similar to the original trade-area model you used to select the site for LITGL's store. You will use this model to evaluate market penetration relative to drive-time distance from that store.

1. Turn off the Trade Area Penetration layer.

2. Click the drop-down arrow in the Business Analyst toolbar, then click Trade Area. Select Create New Trade Area, and click Next. Select Customer Data Required, then click Next. Select Distance Decay Areas, and click Next. Select LITGLStore as the store layer and ID as the ID field. Select the All Stores option. Your screen should resemble the one below. When it does, click Next.

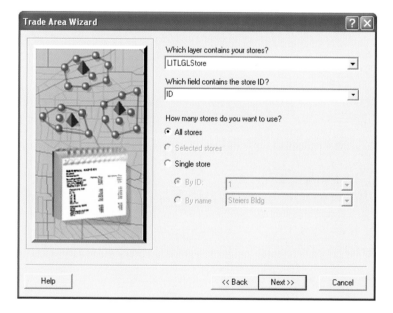

3. In the resulting window, select the Drive Time option, enter **3** as the number of rings you wish to create, and **3** minutes as the size of each ring. Click Next.

4. In the resulting window, select LITGLCustomers as the customer layer, but do *not* select the weight field option. Select StoreID as the Store ID field in the customer layer, then click Next.

5. In the resulting window, select the Calculate using Business Analyst Data option, and select Standard Business Analyst Data (Generalized) as the data source and Current Year Total Households as the data field. Click Next.

6. In the resulting window, select the Generate Report and View report options. Enter **Distance Decay Areas Report** as the Report name and **Distance Decay Areas** as the Trade Area Name. Click Finish.

    Business Analyst Desktop creates the distance-decay trade areas based on drive times and adds them as a layer to the map. Note that these trade areas are comparable in size to the Percent of Sales trade areas, indicating that the Distance Decay trade areas encompass roughly 80 percent of Living in the Green Lane's annual sales, which is significant coverage of the company's sales base.

    Business Analyst Desktop also generates a Distance Decay report, which reports the distribution of customers and penetration rates based on number of customers for the three drive-time trade areas. Note in the report that, as distance from the store increases, the penetration rates of the drive-time trade areas decrease, confirming the central assumption of the distance decay concept.

    You will revise the analysis to determine if this is true for the concentration of purchases across the drive-time areas as well.

7. Click the drop-down arrow in the Business Analyst toolbar, and click Trade Area. Select Manage Existing Trade Area(s), then click Next. In the resulting window, select the Distance Decay Areas, select the Modify Trade Area option, and click Next to revise the Distance Decay Trade Area Wizard by repeating the steps above.

8. Select Customer Data Required, then click Next. Select Distance Decay Areas, and click Next. Select LITGLStore as the store layer and ID as the ID field. Select the All Stores option. In the resulting window, select the Drive Time option, enter **3** as the number of rings you wish to create, and **3** minutes as the size of each ring. Click Next.

9. In the resulting window, select LITGLCustomers as the customer layer. Select the Use a weight in the customer layer option and designate LYPurchase as the weighted field. Select StoreID as field in the Customer layer, then click Next. Select the Standard Business Analyst Data (Generalized) as the data layer and Shelter: Tot as the data field.

10. In the next window, name the report **Distance Decay Areas Sales Report** and accept **Distance Decay Areas by Sales** as the Trade Area name. Complete the wizard and click Finish to run the analysis.

Business Analyst Desktop revises the map layer to reflect these settings and creates a new report. That report resembles the one below. Note that penetration rates measured in terms of sales follows the same pattern as the rates based on number of customers.

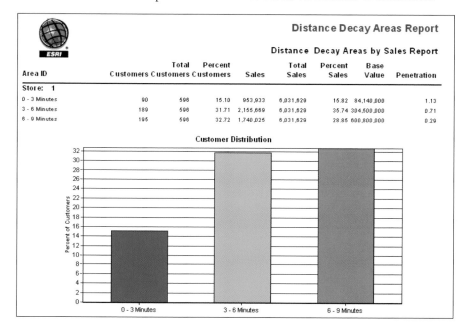

Area ID	Total Customers	Percent Customers	Customers	Sales	Total Sales	Percent Sales	Base Value	Penetration
Store: 1								
0 - 3 Minutes	90	596	15.10	953,933	6,031,629	15.82	84,140,000	1.13
3 - 6 Minutes	189	596	31.71	2,155,669	6,031,629	35.74	304,500,000	0.71
6 - 9 Minutes	195	596	32.72	1,740,025	6,031,629	28.85	600,800,000	0.29

Comparison of the customer profile you created in chapter 6 and the Market Profile and Tapestry Segmentation Area Profile reports from this chapter reveal that customers in the High Purchase segment are similar to the general population in the Percentage of Sales trade areas. This indicates that the full trade areas, not simply isolated customers within them, are attractive prospects for Living in the Green Lane. In addition, the relatively low penetration rates revealed in both the Trade Area Penetration and Distance Decay reports indicate that there is significant potential for sales to new customers within these areas. Based on these comparisons, it is reasonable to conclude that the current LITGL store is a good model to use in selecting locations for future stores. Now you will use the Find Similar tool to locate several potential locations in the Minneapolis-St. Paul area similar to this model and identify the best matches.

Find similar sites with Principal Component Analysis

The Find Similar analytical tool evaluates a set of potential sites to determine which ones best match a site designated as the model store. It uses two approaches: the conventional Find Similar method and a statistical technique known as Principal Component Analysis. In both, you define attributes to be used as a basis for comparison and use them to rate alternative new sites. As the Principal Component Analysis option produces a ranking of all available sites, you will use this approach.

1. Turn off the LITGL Customers layer and all trade-area layers in the map and turn on the Available Properties layer in the Table of Contents. Right-click this layer, then click Zoom to Layer to view all the sites in the layer.

    Note that it displays several potential expansion sites around the Minneapolis-St. Paul area as well as LITGL's store. Janice and Steven's revision of the site criteria they are seeking, as well as their willingness to build new facilities if necessary, serve to increase the range of sites suitable for new LITGL stores. In addition, based on the size of the Percent of Sales Trade Area layer, you will use a 3-mile ring buffer to evaluate these available sites.

2. Click the drop-down arrow in the Business Analyst toolbar, then click Analysis. Select Create New Analysis, and click Next. Select Find Similar, then click Next. In the resulting window, select Available Properties as the trade-area layer, define a 3-mile buffer around each site, and select 11:11 as the master site and Block Groups as the summarization level. The window should resemble the one below. When it does, click Next.

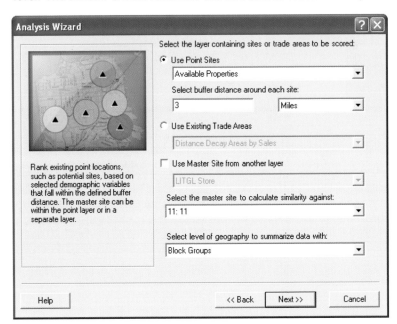

3. In the resulting Window, select Principal Components Analysis as the method to use and enter **6** as the number of sites to rank. Click Next.

4. In the resulting window, select the attributes below from the list of available fields and click the single right arrow to move them to the box on the right. The window should resemble the one below. When it does, click Next.

    CY Owner Occupied HU
    CY Median HH Income
    CY Median Value: Owner HU
    Shelter: Tot

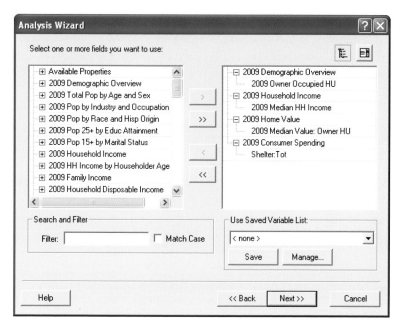

5. In the resulting window, enter **PCA Site Ranking** as the Analysis name, then click Finish.

Business Analyst Desktop performs the principal component analysis you specified, creates a 3-mile buffer for each site, evaluates each using the attributes you specified, ranks the sites, creates a layer which displays the number of ranked sites you specified, and displays them on the map. Your map should resemble the one on the next page.

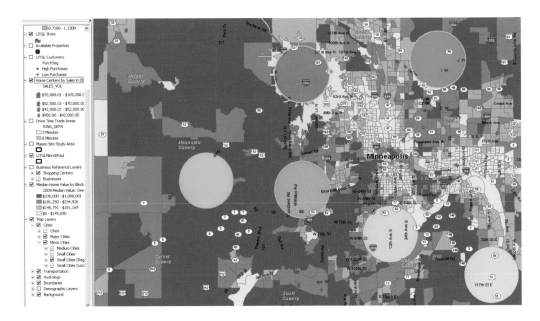

Not surprisingly, the highest ranked site is the model site, LITGL's existing store. All the other sites are those that most closely match this model store. These are the most attractive sites for new LITGL stores.

To view the sites and their rankings, open the attribute table for the new layer. Select the features in the table one by one to view the site on the map and review the values of the site for the attributes used in the selection process. In the table below, the three most highly ranked sites are selected. These sites are those displayed as selected in the map above.

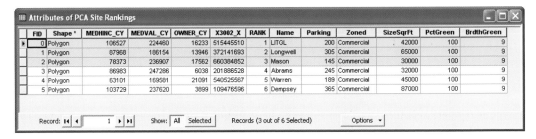

FID	Shape *	MEDHINC_CY	MEDVAL_CY	OWNER_CY	X3002_X	RANK	Name	Parking	Zoned	SizeSqrFt	PctGreen	BrdthGreen
0	Polygon	106527	224460	16233	515445510	1	LITGL	200	Commercial	42000	100	9
1	Polygon	87968	186154	13946	372141693	2	Longwell	305	Commercial	65000	100	9
2	Polygon	78373	236907	17562	660384852	3	Mason	145	Commercial	30000	100	9
3	Polygon	86983	247286	6038	201886528	4	Abrams	245	Commercial	32000	100	9
4	Polygon	63101	169581	21091	540525567	5	Warren	189	Commercial	45000	100	9
5	Polygon	103729	237620	3899	109476596	6	Dempsey	365	Commercial	87000	100	9

These sites have now been ranked on market attractiveness relative to key attributes of LITGL's high value customer profile. You must also evaluate the competitive environment of the most attractive sites in order to recommend the best location. You will use the Advanced Huff Model to do so.

Evaluate site competitiveness with Advanced Huff Model

The Advanced Huff Model allows you to assess the strength of a potential location relative to existing competitors in a general market area by taking into account a variety of factors that affect competitiveness. For example, the attractiveness of a store to a potential customer generally decreases as distance from the store increases. Similarly, attractiveness generally increases as the size of the store increases and, thus, its ability to offer a wide product line. While these two factors influence LITGL's competitive environment, two others are, in Janice and Steven's view, also relevant: the commitment of the store to green-lifestyle products as measured by the percentage of its products that are environmentally friendly, and the breadth of its mix of green products and service. The latter factor is measured as a rating from 1 to 10, with 10 representing a very wide range of green products and 1 a very narrow range.

The Advanced Huff Model allows you to incorporate these factors into your competitive analysis. You will use it to assess the competitive strength of a highly ranked site, a process you would repeat for the other potential sites before making a final selection.

1. Check the box to the left of the Mason Site layer to turn it on. Right-click this layer, click Zoom to Layer to zoom and center the display. Then turn off the PCA Site Ranking layer to get a clear view of the map. Your map should resemble the one below.

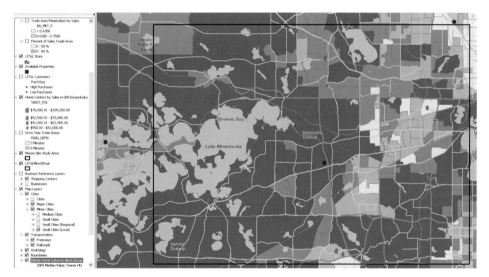

2. If necessary, turn on the Home Centers by Sales in OOO's and Available Properties layers, open their attribute tables, and review the data they include.

    Note that both tables include three attributes with data on the relevant store characteristics discussed by Janice and Steven: store size, percent of green products in the product mix, and breadth of green products. Features differ in size In the Available Properties layer, but have the same values for the green product attributes. These values reflect LITGL's marketing strategy and would be the same for a store at any of the available sites. This is the data you will include in the Advanced Huff Model.

3. Click the drop-down arrow in the Business Analyst toolbar, then click Modeling. Select Create New Modeling Analysis, and click Next. Select Advanced Huff Model with Statistical Calibration, then click Next.

4. In the resulting window, select Block Groups as the Sales Potential layer, ID as the ID field, and Home Imp Material-Own & Rent: Tot as the Sales Potential field. (This attribute reports total expenditures on home improvement materials.) The window should resemble this. When it does, click Next.

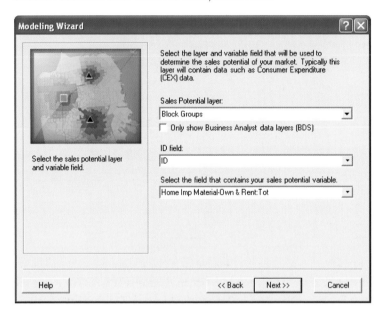

5. In the resulting window, select Home Centers by Sales in 000's as the Competitive Stores layer and ID as the Store ID field, then click Next. In the resulting window, select the By selecting a point feature from layer option, and click Next. In the resulting window, select Available Properties as the potential site layer, and 9: Mason as the feature, then click Next. In the resulting window, select the Enter parameters manually option, then click Next.

6. In the resulting window, select the Use Drive Time option. Add SizeSqrFt, PctGreen, and BrdthGreen to the list of predictor variables using the + key. Assign Coefficients of 1.0, 0.8, and 1.2 to them, respectively. The window should resemble this. When it does, click Next.

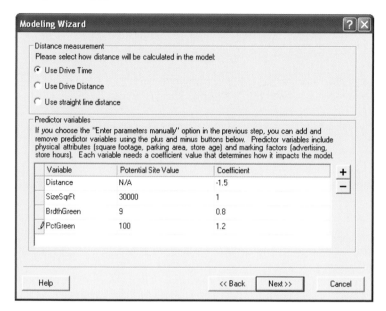

These values reflect Janice and Steven's judgment about the competitiveness factors in this market and their importance. They believe that the percent of green product offerings is a bit more important to consumers than store size, while the breadth of green products and services in a store is slightly less important. Taken together, however, these factors are twice as important as store size and will be major factors in consumer decisions about shopping at a Living in the Green Lane store. The positive values for the three attributes you added indicate a direct relationship, while the negative value for distance indicates an inverse relationship. This is consistent with the judgments discussed above.

7. In the resulting window, enter **Advanced Huff Model for Mason Site** as the trade-area name, then click Finish.

In a process which might take several minutes, Business Analyst Desktop runs the model and displays a box indicating the value of predicted sales in the trade area. Click OK in the box to display the full analysis on the map, which resembles the one on the next page, though it might differ depending upon the extent of the original map screen. The map displays the block groups in the market area with a legend that reflects total projected sales in each based on the balance of attractiveness between this site and competing stores relative to the three factors.

Open the attribute table for this layer and review it. The HM_PROB field reports the estimated probability that residents of a block group will shop at LITGL, a value which may also be understood as the company's projected market share in the block group. The HM_TOTAL attribute reports projected sales in the block group for LITGL. Select the HM_Total attribute and calculate statistics for it. The table and Statistics box should resemble the ones below, though the exact numbers might vary based on your map extent. The Sum figure reports the total projected sales for a LITGL store at this site given the competitive environment.

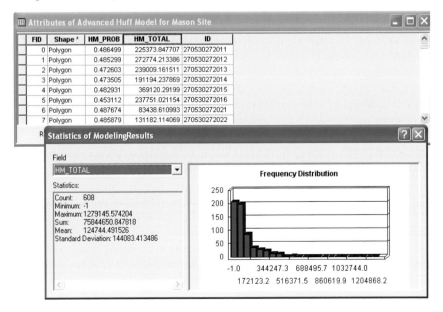

8.   Save this map document in the chapter 7 folder to allow you to return to this analysis.

Repeating this procedure for the other highly rated locations would produce comparative sales projections for them. This data, supplemented with the trade-area analysis procedures you used in chapter 5, allows you to select the two most favorable sites for the second and third Living in the Green Lane stores.

Upon completion of that process, you recommend to Janice and Steven that the Mason and Longwell sites be selected for those stores. They accept your recommendation and select those sites for the company's expansion plans. Coupled with the enhanced merchandising strategy you developed in chapter 6, the company is ready to implement its comprehensive growth strategy and complete its transition from a single green home center to a chain of neighborhood green-lifestyle centers.

ROI considerations

The return on investment equation for site selection in this instance is similar to that presented in chapter 5. For that reason, the focus here will be on return on investment considerations in the product mix and merchandising decisions covered in chapter 6.

In this context, the incremental value of business GIS consists of the differences in revenue between a well-chosen product line addition and a poorly chosen one. Data from the Market Potential Indexes employed at the end of chapter 6 help quantify those differences, which are summarized in table 1 below.

This table summarizes the difference between a product line for which a target segment has a low MPI value and a similar line for which the segment has a high MPI value. The first row reports the number of customers in the High Purchases segment. The balance of the table reports revenue differentials for only these 202 customers. The next row reports average annual purchases across all households for two different products. For comparative purposes, a common value of $1,000 is assumed.

The next row of MPI values for the two product lines are reported. Recall that the value of 80 indicates that segment purchases of this product are 20 percent below average, making this an unattractive product line extension. Conversely, the 120 value in the High MPI column represents an attractive choice. In the next row, these values are used to calculate average annual purchases by each customer in the two segments. Incremental purchases for High MPI customers are reported in the next row and this figure is aggregated for the 202 High Purchasing customers in the final row.

	Low MPI	High MPI
High purchasing customers	202	202
Overall average annual purchases	$1,000	$1,000
Tapestry Segmentation segment MPI	80	120
Segment average annual purchases	$800	$1,200
Incremental purchases per household		$400
Total incremental purchases		$80,800

Table IV.2: Revenue comparisons of new product lines

The figure $80,800, then, estimates the incremental revenue from an attractive product line over a less-attractive product line for these 202 high-value customers. Although this increased revenue figure exceeds the cost of the customer profiling analysis that identifies high-value customers and attractive product lines, it actually understates the full revenue impact of this decision. Several additional factors will increase incremental sales even further. Among them are the following:

1. The table reports values for only 202 high value customers. Sales will also increase from other customers with similar characteristics not included in this group.
2. Well-chosen product lines result in a complementary set of products that are attractive to customers and conveniently available in one store. This combination of assortment and convenience will increase sales across product lines.
3. The expanded assortment of green lifestyle products will attract new green consumers to LITLG's stores, thus increasing the company's customer base and sales.
4. As new and existing customers experience the opportunities of green lifestyle shopping, they will substitute more traditional brand choices with green products from Living in the Green Lane, increasing the sales per customer considerably.

These are the incremental benefits of a customer profiling analysis and its application in merchandising strategy. Customer profiling also can contribute to efficiencies in promotion expenditure by identifying the most used media and most appealing messages for target segments. These efficiencies, coupled with even more substantial sales impact of site selection analysis, produce a revenue stream that far surpasses the costs of Business Analyst Desktop, the Segmentation Module extension, and the research projects that produce these results.

Summary of learning
The analysis in chapters 6 and 7 has enabled you to create a new business model for Living in the Green Lane, develop merchandising and promotional strategies to support it, and select the best sites for the company's continuing retail expansion. In that process, you have

expanded your business GIS knowledge and skill set. Consider for a moment what you have accomplished. You have learned:

1. The value of integrating internal customer data into GIS projects
2. The benefit of identifying high value customers
3. The value of creating geodemographic and lifestyle profiles of high value customers
4. The value of Market Potential Indexes in understanding the values, preferences, and buying patterns of target customers
5. The benefit of matching marketing strategies with target customer profiles
6. The value of using customer purchasing data to create trade areas
7. The benefit of using market penetration and distance decay to measure store attractiveness
8. The value of ranking the attractiveness of available sites and estimating market penetration for a new store

Your have enhanced your GIS skills by using Business Analyst Desktop to:

1. Geocode customer data with locator services
2. Use Layer Properties to view attribute distribution and identify high volume customers
3. Use a spatial join to attach demographic and Tapestry Segmentation attributes to customer features based on their location
4. Use summary tables to calculate geodemographic and Tapestry Segmentation lifestyle profiles of high volume customers
5. Use Tapestry Segmentation data with Market Potential Indexes to identify customer values, media habits, product preferences, and purchasing patterns
6. Use this information to make product line and merchandising decisions appropriate for Living in the Green Lane's best customers
7. Create sales-derived trade areas from customer records
8. Produce trade area penetration and distance decay reports
9. Rank available new sites using Principal Components Analysis Estimate market penetration and sales using Advanced Huff Model Analysis

Notes
1. We request a bit of chronological indulgence here. All the data provided with this book is the most recent available at publication. The Business Analyst Online data is the most recent, period. Clearly then, there is no two-year gap between the data used in previous chapters and that used from this point forward. We ask you to overlook that chronological inconsistency for the sake of the Living in the Green Lane story line.

2. Kotler, Phillip and Kevin Keller. 2008. *Marketing management.* New York: Prentice Hall.

3. Hughes, Arthur. 2005. *Strategic database marketing 3rd edition.* New York: McGraw Hill.

4. Buttle, Francis. 2008. *Customer relationship management.* Boston: Butterworth-Heinemann.

5. The Spatial Overlay Analysis Procedure in Business Analyst also provides this capability when used with standardized BA data. The spatial join technique is used here as it can employ the calculated fields in the customized shapefile upon which the Median Home Value by Block Group layer is based.

6. These figures are derived from the *2007 Tapestry Segmentation Summary Table*, which is available as a PDF file as part of the Business Analyst Segmentation Module or from the Tapestry Segmentation Demonstration CD.

Part V

Sales territory management and route optimization

Relevance	Businesses wishing to design and/or revise sales territories. Businesses with external sales, service, maintenance, and/or customer support operations that wish to reduce operating costs by optimizing routing efficiency.
Business scenario	As part of its green-lifestyle initiative, LITGL has added organic lawn and garden and pest-control services to its product line. The company will use a direct-sales staff to call on potential commercial and residential customers to market these services and customize them to each client's needs. It will deliver the services with a fleet of service vehicles and personnel, each of whom covers clients within defined territories.
Analysis required	LITGL must create sales territories for each of its direct-sales representatives. The territories should generally be balanced in size and sales potential. The company also must provide its staff with the most efficient routing information for their daily customer service calls.
Role of business GIS in analysis	Create Territories layer using Territory Design extension. Build sales territories around sales-staff locations relative to LITGL's three stores. Organize sales territories by store. Balance territories relative to sales potential. Identify clients to be included in customer service technician's daily route. Design the most efficient routes for daily service calls. Determine resulting efficiencies and savings.
Integrated business GIS tool	ESRI Business Analyst Desktop, Territory Design extension.
ROI considerations: cost of business GIS	Business Analyst Desktop purchase, salary of GIS analyst.
ROI considerations: benefits of business GIS	Increased sales from well-balanced and efficiently serviced sales territories. Decreased costs and increased efficiency and customer satisfaction from optimizing the efficiency of service routes.
Environmental impact of business decision	Lower operating costs and fewer car emissions from compact, efficient sales territories. Lower driving time and costs as well as more efficient operations for customer service fleet.

Table V.1 Executive summary

The Living in the Green Lane scenario

As part of its evolution from a green-home center to a green-lifestyle center, Living in the Green Lane enhanced its product line by adding, among other things, organic lawn-and-garden-care services and organic pest-control services. As organic approaches involve managing lawn/garden health and pest control simultaneously, the company will emphasize integrated service packages. However, customers also may purchase contracts separately for the specific services they desire.

To promote these services, the company will add a direct-sales force to open and manage residential and commercial accounts. To deliver services, LITGL will use a fleet of vehicles staffed by customer-service technicians. To optimize the profit potential and minimize the environmental impact of these initiatives, the company must design an efficient system of sales territories and manage the logistics of service delivery effectively.

Each Living in the Green Lane store in its initial implementation will employ two sales representatives to serve households in assigned territories near each store. Consistent with its philosophy of low-impact business operations, establishing a local presence in each market area, and providing high-quality customer service, LITGL sales reps will live in the neighborhoods they serve.

This approach requires that the company design a system of six sales territories—two for each of its three stores. The territories should be reasonably balanced in terms of geographic size, number of households, and sales potential as measured by annual purchases of lawn-and-garden maintenance services.

The role of the sales staff is to establish new accounts for LITGL's services, maintain high levels of satisfaction among existing customers, and exploit emerging opportunities for cross-selling of enhanced services to new and existing customers. Thus, sales reps will continue to manage customer relations when household accounts have been established, but will not be responsible for service delivery, which will be assigned to teams of lawn-care and pest-control technicians.

Initially, Living in the Green Lane will employ one lawn-service team and one pest-control service technician per store, with each team and technician servicing accounts in two sales territories. As the business grows, additional service personnel will be added and aligned with sales territories. The goal is to ensure that the sales representative and the service team for each territory coordinate their efforts to serve local customers responsively and effectively.

To implement this program, Living in the Green Lane must design a system of balanced territories for sales representatives and a routing system that maximizes the efficiency of service personnel teams. You will employ ESRI Business Analyst Desktop's Territory Design system and Find Route functionality to achieve these goals.

Integrated business GIS tools in sales territory design

Territory design is a significant element in the sales-management process. Its objective from an operational level is to maximize the efficiency and effectiveness of the sales force. However, it also has a substantial impact on both sales-force motivation and customer satisfaction. This can be especially true in territory realignment decisions, which often call additional interpersonal and professional dynamics into play. Integrated business GIS is a useful tool in achieving the best balance among these often-conflicting dynamics.

From the perspective of operational efficiency, the objective of territory design systems is to balance sales potential, workload, and account service quality. The ideal is a set of geographically compact territories that may be served efficiently and that provide similar levels of sales potential measured in terms of total purchases and/or existing or potential sales accounts.

The benefits of a well-designed territory system are manifold. Compact territories allow representatives to serve them efficiently, maximizing the ratio of selling time to travel time. Equalizing the sales potential of territories serves the objectives of sales force morale and customer satisfaction. Representatives with balanced territories have comparable opportunities for high levels of performance—especially important when their compensation is based in whole or in part on sales levels. Requisite levels of customer service in balanced territories can be maintained across the sales organization, preserving the strength of important customer relationships. Given so many factors to consider, perfectly balanced territories are nearly impossible. This places a premium on systems which provide flexibility in assessing and adjusting alternative territory structures.

To achieve these goals, sales managers must define the attributes used to measure territory potential and, therefore, to determine territory balance. These measures in established sales organizations include number of current and potential accounts, size of those accounts, and estimates of aggregate demand within each territory.

These measures in business-to-business settings emerge from the analysis of potential business customers in each territory, their size, and the role of the selling company's products in their operations. These measures in business-to-consumer settings are related to the number of current and potential customer households, and their average annual expenditures in the company's product line. Further, a mix of existing customers to provide reliable revenue and new prospects to provide opportunities for sales growth is desirable. Customer relationship management systems in sophisticated sales organizations track this information and use it to nurture a comprehensive interaction between the entire company, not simply the sales function, and its customers.

Detailed customer information is unavailable in "new-to-market" sales situations—exactly the challenge Living in the Green Lane faces. Sales managers in this context must use other measures to estimate sales potential. Business-to-business settings call for identifying potential customer organizations, understanding their use of the selling company's offerings, and estimating annual expenditures on these goods and services. For these tasks, the business listings functions of Business Analyst Desktop offer a valuable starting point for estimating market potential and balancing it across territories.

In business-to-consumer settings—also a situation LITGL faces—sales potential is a function of the number of potential customers, which usually is measured in population or households, their average annual expenditures on relevant products, and the resulting aggregate demand for these products in each territory. Business Analyst Desktop supports this effort with its comprehensive collection of current household demographic and consumer expenditure data at a range of geographic levels.

Your task, then, is to use the Territory Design extension of Business Analyst Desktop to develop a territory system for Living in the Green Lane's new-to-market, business-to-consumer personal sales force. To implement Janice and Steven's vision of the sales function, you will create two sales territories for each of the company's three stores in the Minneapolis-St. Paul area. Specifically, you will identify two sales representative locations, or seed points, for each store, specify the attributes you will use to determine sales potential, and use Territory Design to create appropriate territories by combining ZIP Codes based on the specified characteristics.

Once the territory system is developed, you will examine its balance using the data exploration tools of Territory Design. If necessary, you will realign the default territories by moving ZIP Codes between them to rebalance their sales potential measures and create more serviceable territories.

Although this will complete your analysis, be sure to note as you use Territory Design that it also is capable of analyzing sales potential measured by the number and size of target businesses. For more mature organizations, it is also capable of integrating information from current customer accounts into the analysis as well. These capabilities are very useful in realigning a mature sales territory scheme to better match emerging demand trends in the service region.

Integrated business GIS tools in route optimization applications

Territory design is very important, yet it is simply the initial step in managing sales and service organizations. Additional tasks in this process include maximizing efficiency by optimizing the daily travel routes of sales and/or service personnel. Efficient routing reduces travel time relative to sales/service time, lowers service fleet costs for fuel and maintenance, and increases revenues by allowing sales/service personnel to make more calls daily. Indeed, as you can see below, optimized routing applications offer one of the clearest, most predictable, and easily achievable sources of ROI among business GIS applications.

The Find Route tool in Business Analyst Desktop provides the capability of route optimization. Although this technology is relevant to daily calls for sales representatives, you will use it to determine the best route for a service technician providing pest control services. Initially, each store will deploy one pest-control service technician and one lawn-and-garden maintenance team to serve customers spread across its two sales territories. As growth warrants, additional technicians will be added to serve the customer bases and service territories, which will be realigned as necessary.

Specifically, you will use Business Analyst Desktop's routing function to determine the optimum route for one service technician's daily calls. You then will calculate the distance and time saved by this route and determine its advantages in lower costs and increased sales calls.

While this will conclude your routing application, consider for a moment the potential benefits of extending your analysis to cover more complex situations. One major enhancement is the ability to consider the differing time and technical demands of each call in service fleet scheduling. In LITGL's case, routine pest control calls will be less time consuming than initial system installations, while lawn-and-garden maintenance calls will vary in length depending upon the size of customers' yards and the services they select. Including this information in routing systems increases their accuracy and predictability and, therefore, increases service levels and customer satisfaction.

A second major enhancement of routing systems is real-time integration of information flows between service technicians and dispatchers. Service vehicles with GPS capability allow real-time tracking of daily routes as they proceed. Further, by integrating real-time traffic information into the system, dispatchers and technicians can adjust planned routes as driving situations develop. These systems offer the additional benefit of accommodating calls for emergency service into the system as they arise. Dispatchers can assign the nearest technician—or the technician with the most available time—to these calls, reducing their impact on the overall efficiency of the original route plans.

Thus, chapter 8 will teach you how to use Business Analyst Desktop tools that offer significant contributions to the design and management of efficient, effective sales organizations.

Chapter 8

Sales territory design and balancing; route optimization

ESRI Business Analyst Desktop supports sales management and customer service functions through the Territory Design extension, which builds well-balanced sales territories, and the Find Route tool, which enables route optimization for service personnel in each territory. As the latter function assumes a functioning sales territory system, you will begin by creating such a system with the Territory Design extension.

Run Business Analyst Desktop; load map

1. Click Start, Programs, ArcGIS, Business Analyst, BusinessAnalyst.mxd to run ArcMap, load the Business Analyst Extension, and then load the default Business Analyst map.

2. Click OK when the Update Spatial Reference dialog box appears, then close the Business Analyst Assistant window on the right of the screen.

3. Click File, then click Open. Navigate to C:\My Output Data\Projects\LITGL Minneapolis St Paul\CustomData\ChapterFiles\Chapter8\LITGLSalesMgt.mxd. Click the map file to open it.

This map contains a layer displaying available Living in the Green Lane's Steiers, Mason, and Longwell stores over a thematic map of Home Related Expenditures per Household by ZIP Code. The data source for this layer is a shapefile similar to the one you created in Chapter 3, but based on ZIP Codes. ZIP Codes are a more useful geographic unit than block groups for defining sales territories. The Table of Contents also includes a Sales Reps' Homes layer that is turned off. These are the geographic points around which you will build territories.

Load Territory Design and create territories layer

1. From the main menu bar, click Tools, then Extensions to open the Extensions dialog box. Select Territory Design at the bottom of this list. Click the Close button.

2. Click View, then Toolbars to open the Toolbars dialog box. Check Territory Design near the bottom of the list to open the Territory Design toolbar. Drag and drop it to the top of the screen to dock it there.

   With the Territory Design toolbar in place, you are ready to begin the process of territory creation. The first step is to create territories layers in the current map. In this case, you will create layers for the two levels of sales organization: sales territories and stores. You will have two sales territories for each of Living in the Green Lane's three stores. Using Territory Design, you will create territories around six seed points—the offices of the six local sales people who will represent LITGL's organic pest-control and lawn-and-garden maintenance services. Once the territories are created, you will examine their characteristics and adjust their composition to achieve a balance of sales potential across them.

3. Click the drop-down box on the Territory Design toolbar, click Active Layer, then click Create Territory Layer. (You may also click the Create Territory Layer icon )

4. In the resulting window, select ZIP Codes, ID as the ID field and 2 as the Number of territory levels. Specify that you wish to create 6 Territories in 3 Regions, which you should rename to **Stores**. When your window resembles the one below, click Next.

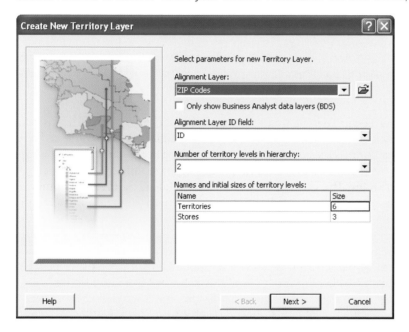

5. In the resulting window, select the following BA Fields as Setup variables for the new layer levels and move them to the right window as Selected variable(s).

General Count
CY Total Population
CY Total Households
CY Owner Occupied HU
CY Pop 25+ by Educ: Assoc Deg
CY Pop 25+ by Educ: Bach Deg
CY Pop 25+ by Educ: Grad Deg
CY Median HH Income
CY Median Value: Owner HU
Lawn & Garden: Tot

When the window resembles the one below, click Next.

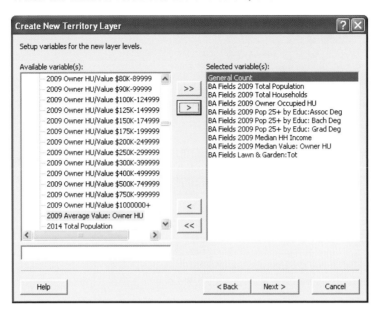

6.   In the resulting window, enter LITGL **Sales Territories** as the New Territory Layer Name, then click Next. Review the settings and click Finish.

Territory Design creates two new territory layers, Territories and Stores, and populates them with 6 and 3 features each, matching your specifications. Though turned on, neither layer is displayed on the map as you have not yet assigned geographic units to either of them. Open the attribute tables for each. Each table contains the attribute columns you specified, but has no data. These data fields will be populated when you add geographic units to each Territory and Store in the next phase of the territory design process. The system also creates a ZIP Code alignment layer for use in balancing territories, which is turned on.

Creating territories

You will now create territories around selected geographical points by aggregating ZIP Codes around those points. In this case, the starting points are the homes of the local sales representatives in neighborhoods around Living in the Green Lane's three stores. Territory Design will use these locations as seed points for the territory system. You will balance the territory system based on distance, population and the number of owner-occupied housing units.

1.   Click the Territory Design drop-down menu, then Create Territories (or click the Create Territories icon ![icon]) to open the Create New Territories wizard. In the first window, select LITGL Sales Territories as the Territory Layer. Select the Create territories from locations option, then click Next.

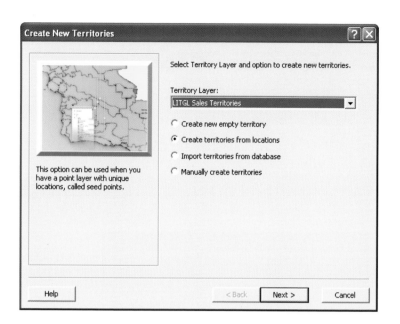

2. In the resulting window, select Territories as the Territory level to create, select the Remove and replace existing territories option, designate SalesRepHomes as the point layer, Name as the Attribute for naming territories, and enter **6** as the Number of territories to be created. When your window resembles the one below, click Next.

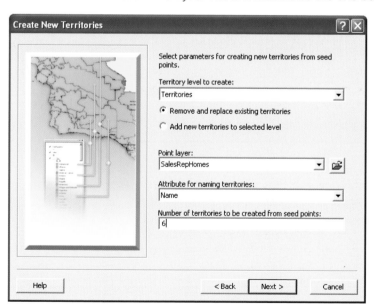

3. In the resulting window, designate Straight line distance as the Distance type, Miles as Distance units, enter **20** as the Maximum distance a territory lies from its seed point, and select Centroid of geography element as the Distance measurement method. (Note: the 20-mile limit here is designed to produce compact territories in the interest of local service. To expand the service area, LITGL would opt for additional territories rather than expanding existing ones.) Click Next.

4. In the resulting window, select the Balance territories option, which will balance the initial territory system using attributes you select. Click Next.

5. In the Balancing Options window, select CY Total Population and CY Owner Occupied HU as the Balancing variables. When your window resembles the one below, click Next.

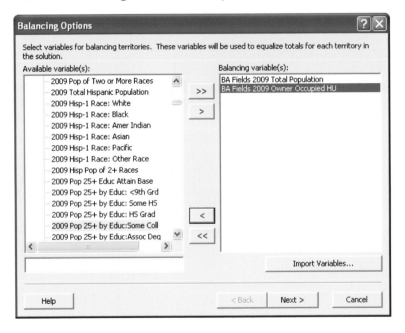

6. In the resulting window, you will assign weights to the two balancing variables you selected and to a third factor, Distance. You may do so by entering values in the table or by using your mouse to adjust the boundary lines between factors in the pie chart at the bottom of the window. If you choose the former method, be sure to enter values which sum to 100. You may adjust the balance of your territory system later, but for now, use initial weights of 50 for CY Owner Occupied HU and 20 for Distance, and 30 for CY Total Population. CY Owner Occupied HU is weighted heavily as it is the best measure of potential customer units for the services LITGL will be offering. When your window resembles the one on the next page, click Next.

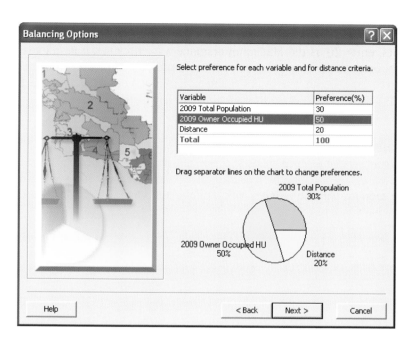

7. In the resulting window, select both options, then click Next. Review your settings in the resulting window and, when you are satisfied, click Finish.

Territory Design applies your settings and creates a territory system using your specifications. Specifically, it aggregates ZIP Codes into sales territories around the points in the Sales Reps' Homes layer, balances the territories using the variables and weights you designated, calculates the attributes you selected for each territory, populates the attribute table of the sales territories layer with those values, and displays the resulting territories on the map. These are complex calculations and may take some time. When they are complete, your map should resemble the one below, though the colors may differ.

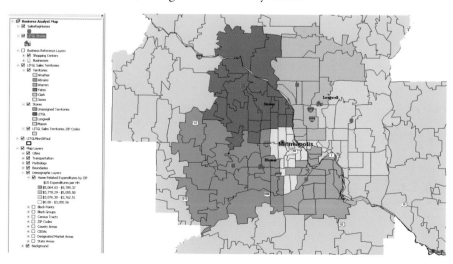

Open the attribute table for the layer to compare the values of the territories for the attributes you designated, including two attributes which report the total number of households and expenditures for lawn and garden services in each territory. You will revise and rebalance the territories in this system later. Before moving to this step, however, you will create regions around the three LITGL stores, each of which contains two territories.

Creating regions around stores

Each of Living in the Green Lane's three stores in your territory system supports two sales territories. You will create this association by repeating the Create Territories procedure using the LITGL Stores layer to supply seed points.

1. Click the Territory Design drop-down menu, then Create Territories (or click the Create Territories icon ▰) to open the Create New Territories wizard.

   In the first window, select LITGL Sales Territories as the Territory Layer. Select the Create territories from locations option, then click Next.

2. In the resulting window, select Stores as the Territory level to create, select the Remove and replace existing territories option, designate LITGL Stores as the point layer, Name as the Attribute for naming territories, and enter **3** as the Number of territories to be created. When your window resembles the one below, click Next.

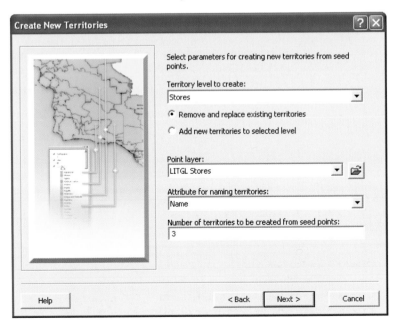

3. In the resulting window, designate Straight line distance as the Distance type, Miles as Distance units, enter **40** as the Maximum distance a territory lies from its seed point, and select Centroid of geography element as the Distance measurement method.

(Note: the 40-mile limit here is designed to ensure that all sales territories are included within the sales regions you are creating for each store.) Click Next.

4. Complete the remaining windows in the wizard, specifying the same settings values you designated when defining territories. In the Balancing Options box, specify that territories should be contiguous and have no holes. These entries ensure that the settings for the store region level will match those of the territories level. In the final window, review your settings, then click Finish.

    Territory Design repeats the calculation process you just performed at the territory level and displays the results on the map. You may not see the boundaries of the defined regions until you turn off the Territories layer. When the Stores layer is displayed, your map should resemble this one, though the colors may differ.

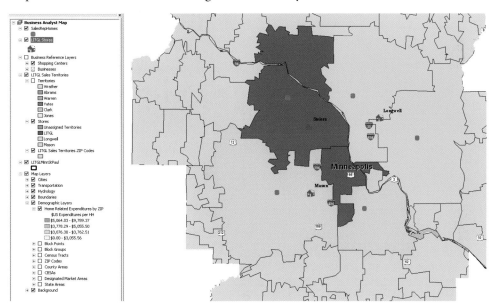

5. With the Territories layer displayed, click the Edit Mode button on the Territory Design toolbar. Note the effect on the map. The Territories layer is turned off, and the boundaries of the Stores region layer are displayed in bold. This allows you to view the relationship between the Stores and Territories layer very clearly. Click the Edit Territories button again to turn it off.

Refining a territory system statistically

The next step in the territory design process is to review the initial system produced by Territory Design and refine it to meet LITGL's objectives more closely. The Territory Design Window and Edit process allow you to assess the current balance among territories and move ZIP Codes from one territory to another to improve the overall structure.

1. Click the Territory Design toolbar and click the Show/Hide Territory Design Window icon to open this window at the bottom of the screen. Drag the window to the bottom horizontally if it opens vertically next to the Table of Contents. If necessary, adjust the boundary of the window so it is visible. Use the drop-down box for the Territories field at the top left of this Territory Design Window to display the values for each of the six territories. Your screen should resemble this.

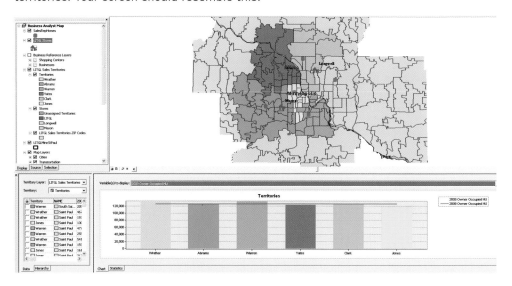

   This window provides a rich environment for exploring and revising your territory structure. Take a moment to review some of its main features.

2. Review the bar charts in the main window. These represent the values of each territory for the most significant balancing attribute, in this case CY Owner Occupied HU. The horizontal line displays the average value across territories, with the bars depicting the values of each of the individual territories. You may adjust the variable displayed in the chart by selecting from available options using the drop-down arrow at the far right of the Variable(s) to display field. This allows you to compare territories on all the attributes you selected when you created the territory layer.

3. At the bottom of the main window, note the Chart and Statistics tabs. The Chart tab is currently selected. Click the Statistics tab to view descriptive statistics for all the variables in the Territories layer. Click Chart when you are finished to return to the Chart view.

4. You may select individual ZIP Codes within territories and view their characteristics in the data box at the lower left of the window. Note that each ZIP Code is identified with the color and name of the territory to which it is assigned. Click a ZIP Code to select it and note that it is highlighted on the map as well. You may use Shift-click and Ctrl-click to select multiple ZIP Codes. Scroll horizontally through this table to view the attributes of each ZIP Code for all territories.

5. You may lock a ZIP Code by checking the box in the leftmost column of its record. This means that it cannot be reassigned to another territory until it is unlocked. When you lock a ZIP Code, a new layer, Locked Territories, appears in the LITGL Sales Territories group layer. In addition, the locked ZIP is displayed with a cross hatch pattern on the map. (This might be hard to see depending on the color of the territory, but it will be visible if you zoom to the Locked Territories layer by right-clicking the layer and selecting Zoom To Layer.) This feature allows you to designate those features that should not be reassigned in the process of refining and rebalancing territories.

6. You may change the level of the territory system that is displayed by selecting the layer of your choice in the drop-down menu for the Territory field. Select the Stores layer to view the values of the layer attributes for each of LITGL's stores and the two sales territories assigned to each.

    When you have finished your review of these functions, select the Territories layer in the Territory field to return to your original view.

    Review the statistics for the territories. They are reasonably well balanced relative to the key attribute of owner-occupied housing units.

    From this perspective, this territory system is acceptable. You also wish to assess territory balance as measured by purchases of lawn and garden maintenance services, which is measured by the attribute Lawn & Garden: Tot.

7. In the Variable(s) to display dropdown box, select Lawn & Garden: Tot and review the resulting graph.

Relative to this measure, the territories could be balanced more favorably. This will be one of your objectives as you refine the territory system to correct for this as well as spatial problems that must be resolved.

8

Refining a territory system spatially

In evaluating a territory system spatially, you are looking for ZIP Code assignments that are inconsistent with store distribution patterns, transportation networks, or travel efficiency. The Territory Design Window allows you to reassign ZIP Codes among territories to overcome these problems.

1. With the Territory Design Window visible, zoom to the extent shown in the map below. If necessary, and turn on the Territories layer.

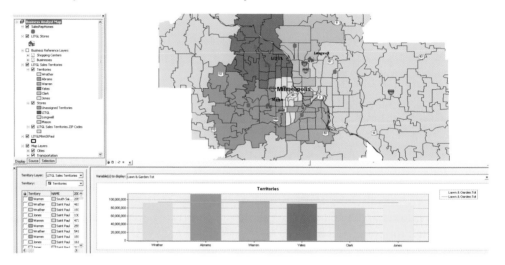

At this scale, several problems appear in the territory structure. Two sales reps, Jones and Warren, have small sections on the opposite side of the Mississippi River from their homes and the majority of their territory. Wrather, another rep, has territory that includes some inconvenient ZIP Codes that would require traversing Warren and Jones' territories to serve.

Look at the left bottom portion of Jones' territory. Note how close to the Mason store these ZIP Codes are and how inconvenient it would be for Jones to serve them. Turn off the Territories layer and, if necessary, turn on the Stores layer. Note that Jones' territory is assigned to the Steiers store. It is inappropriate for this territory to include ZIP Codes so close to the Mason store.

To remedy each of these problems, your will select ZIP Codes and reassign them to other territories.

2. Click the Select Tool 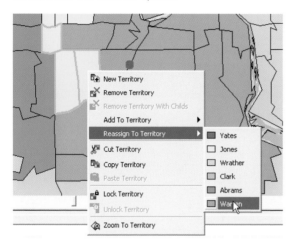 at the far right of the Territory Design toolbar to activate it. Press the Shift key and click in the three ZIP Codes immediately adjacent to the Mason store to select them. Right-click and select Reassign To Territory, then select Warren's territory from the available options. Your screen should resemble this. When it does, click the Warren option.

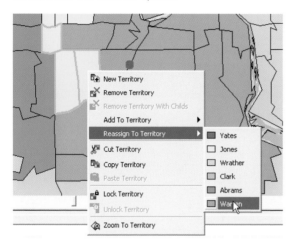

Territory Design reassigns the ZIP Codes to their new territory and recalculates the measures for the territories in the Territory Design window. Note that the Warren territory is now above the average for Lawn & Garden: Tot, while the Jones territory is now well below it. You will correct the problems identified above, then reassign other ZIP Codes to rebalance the sales potential for the territories as measured by Lawn & Garden: Tot.

If you choose, you may also use this process to add currently unassigned ZIP Codes to a sales territory. To do so, click on the ZIP Code to select it, right-click, select Add To Territory, then click the name of the territory to which you wish to add the ZIP Code.

Continue this process until you have resolved the problems defined above and achieved a reasonable balance of sales potential among the six territories. Select the Stores level as well to confirm that your solution also balances sales potential reasonably well among the three stores.

The screen on next page illustrates a favorable resolution of these problems and a reasonable distribution of sales potential across the territories. Your map should resemble this one, though territory colors and the precise allocation of ZIP Codes to territories might differ.

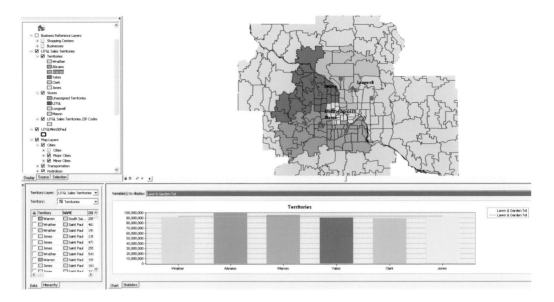

You have completed the process of realigning and rebalancing territories and will now produce two reports to summarize and communicate the results of your work.

Producing territory system reports

Territory system reports present information on each of the territories you have designed at the Territories and Stores level, as well as comparative information on different territory schemes. You will produce the general report first.

1.   Click the drop-down arrow on the Territory Design toolbar, click Reports, then click Territory Report to open the Territory Report dialog box. Enter LITGL **Sales Territories Report** as the report title. When the box resembles the one on next page, click OK.

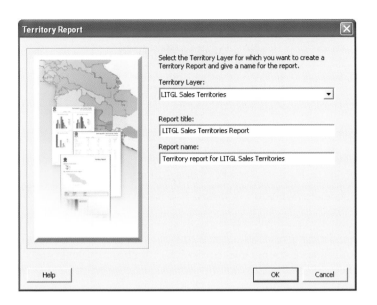

Territory Design creates the report and displays it in a new window. The first page of the report should resemble the one below. Review the content of the report, which presents maps, summary characteristics, descriptive statistics and charts at the Territories, Stores, and Area levels. The Area level includes aggregate data for the entire geographic area included in the territory structure. The Report Window includes options to print the report or export it in a variety of data formats.

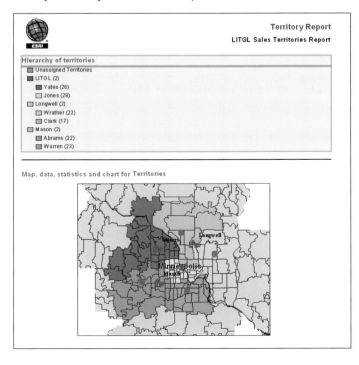

As you experiment with the territory design process, you may wish to compare the characteristics of different territory structures. The Compare Territory Solutions Report function in Territory Design provides the opportunity to do so.

To compare territory design solutions, you must have at least two different territory layers. As you have generated only one, you will add a second territory layer to the Table of Contents.

2. Click the Add Data icon ✚ on the main toolbar. Navigate to C:\My Output Data\ Projects\LITGL Minneapolis St Paul\Territories\PopBalTerritories\tdlayer.lyr. Double-click this layer file to add it to the Table of Contents and the map.

    Review the PopBalTerritories system for a moment. It was generated using CY Total Population as the major balancing variable. That is, this system generates territories with roughly the same total number of people in each. Use the functions of the Territory Design Window to explore the characteristics of this system. Examine the charts for CY Total Population as well as the other layer attributes to see how they are distributed across territories. Note especially the variation in sales potential in the Lawn & Garden: Tot attribute across territories. In the Statistics window, compare the Min(imum) and Max(imum) values for this measure to those in your first territory scheme.

    The Compare Territory Solutions Report is designed to facilitate this type of comparison. It is invaluable in the process of choosing between alternative territory schemes, each of which emphasizes selected objectives at the expense of others. You will now generate that report.

3. Use the Show/Hide Territory Design Window to turn the window off.

4. Click the drop-down arrow on the Territory Design toolbar, click Reports, then click Compare Territory Solutions Report to open the Compare Territory Design Solutions Report dialog box. Select LITGL Sales Territories as the First Territory Layer and PopBalTerritories as the Second Territory Layer. Accept the default title and report names. When the box resembles the one on the next page, click OK.

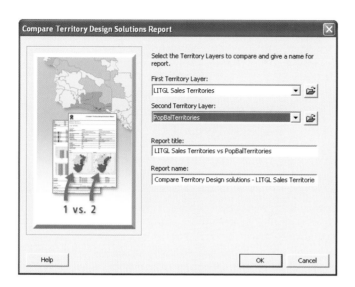

Territory Design performs a comprehensive comparison of the two territory schemes and creates a report which presents this information with maps, charts, tables, and data. The first page of the report resembles the one below. Comparative information on all layer attributes is provided by territory and by store.

Compare Territory Design Solutions Report

LITGL Sales Territories vs PopBalTerritories

Solution1: LITGL Sales Territories
Solution2: PopBalTerritories

Level: Territories

Parameters

Parameter	Solution1	Solution2
Territories creation type	From Locations	From Locations
Manual correction	Yes	No
Create from locations options		
Creation mode	Remove and replace existing territories	Remove and replace existing territories
Seed points layer	C:\My Output Data\Projects\LITGL Minneapolis St Paul\Custom Data\MapFiles\Ch8SalesRepHomes	C:\My Output Data\Projects\LITGL Minneapolis St Paul\Custom Data\MapFiles\Ch8SalesRepHomes
Distance type	Straight line distance	Straight line distance
Distance units	Miles	Miles
Maximum distance that a territory is away from its seed point or center	20.00	20.00
Distance is measured to	Centroid of geography element	Centroid of geography element
Preferences for each variable and for distance criteria	Variable: 2009 Total Population, Preference: 30.00 Variable: 2009 Owner Occupied HU, Preference: 50.00 Variable: Distance, Preference: 20.00	Variable: 2009 Total Population, Preference: 60.00 Variable: Distance, Preference: 40.00
Territories must be contiguous	Yes	Yes
No holes inside territories	Yes	Yes
Attribute for naming territories	Name	Name
Number of territories to be created from seed points	6	6

This report facilitates the territory design process by supporting detailed comparison of alternative territory systems. For example, review the figures for Lawn & Garden: Tot for the two territory schemes on page 3 of the report. In the territory scheme you designed, the sales potential of the lowest territory is approximately 89 percent that of the highest territory, depending upon your specific ZIP Code assignments. In the second territory system, which was balanced by population, the sales potential of the lowest territory is only 63 percent that of the highest territory. This discrepancy would adversely affect both sales force morale and sales call efficiency, both of which would reduce sales force performance. Territory Design in general, and this report specifically, allow you to minimize this type of impact.

You have completed the territory design process with the Territory Design extension of Business Analyst Desktop. You will now turn your attention to the task of optimizing routing within an established service territory.

5.   Save the current map as **LITGLSalesMgtTerr1.mxd in** C:\My Output Data\Projects\LITGL Minneapolis St Paul\ChapterFiles\Chapter8\. Close the Territory Design toolbar in preparation for the next task.

Optimizing routing efficiency of service personnel

Living in the Green Lane's personal selling programs for pest control, as well as lawn-and-garden-maintenance services, have proven successful. The company must now service its new customers. To do so, it has employed a lawn-and-garden-maintenance team and a pest-control service person at each store. They each serve customers in the store's two sales territories. Additional service personnel will be added as the customer base increases.

Lawn and garden services are time consuming and the maintenance technician makes relatively few calls per day. However, the pest-control technician can service each site with a monthly check or a quarterly replenishment of pest control stations very quickly. Thus, these technicians can make many service calls per day and driving time is a significant factor in their efficiency.

To minimize the drive time of these technicians and maximize the number of service calls they make each day, you will use Business Analyst Desktop's routing tools to determine the optimum route for a pest control technician's daily service calls.

1.   Click File, then click Open. Navigate to C:\My Output Data\Projects\LITGL Minneapolis St Paul\CustomData\ChapterFiles\Chapter8\LITGLSalesMgtRoute.mxd. Click the map file to open it.

The Table of Contents includes layers for LITGL's stores and a new layer, Service Calls, which contains information on the Longwell store's pest-control service calls for one day. These are listed in the order they were added to the schedule by the store's dispatcher.

2. Click the drop-down arrow on the Business Analyst toolbar, click Tools, then Find Route to open the Find Route dialog box.

3. With the dialog box open to the Define Stops tab, click the Add Stop button ![add stop icon] to select it. With the Add Stops tool active, click on LITGL's Longwell store on the map to add it as the first stop. The address of the store appears and, because it depends on the exact point you clicked, may not match the exact address in the box below. Select the Return To: option at the bottom of the box to indicate that you want the route to return to this store at its conclusion. The address for this location now appears in the Return To field.

4. Click the Get Points button, then click on the Get Point Feature(s) option. Use the drop-down menu to select Service Calls as the point feature to open. When the box resembles the one on the next page, click the Add Stops button.

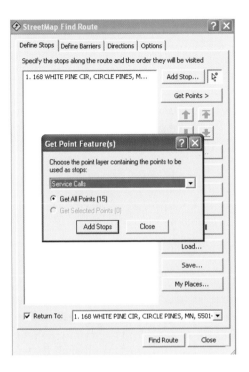

5. Click the Find Route Button. Business Analyst Desktop calculates the route between the stops in the order they are listed, displays the results on the map, and creates a set of directions for the route, which it displays in the Directions tab. Click this tab to view those directions. At the top of the directions box, you will find a summary of the total mileage and drive time of the route. Specific directions are included in the box itself. Your screen should resemble the one below. (Note: If the callout boxes do not appear for each stop, click the Options tab, then select the Callout option at the top of the Options window.

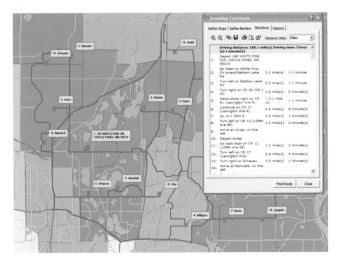

This is the route the technician should drive if the stops are listed in order of entry into the dispatch system. However, as you can see on the map, it is an inefficient route. You wish to reorder the service calls to optimize the efficiency of the route. You will do so with commands on the Options tab.)

6. Click the Options tab. About two-thirds of the way down the box, locate and select the Reorder stops to find optimal route option. Review the other options available in this box, as they provide additional customization options for your analysis. When the Options box resembles this, click Find Route.

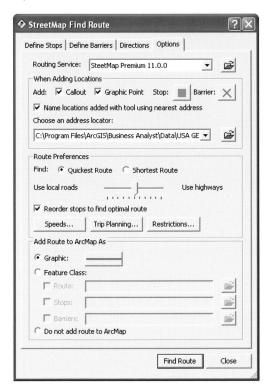

Business Analyst Desktop recalculates the route among the designated stops, but this time it reorders the stops to produce the most efficient route. It adds this route to the map and revises the Directions tab to reflect the new route. Your map should resemble the one on the next page.

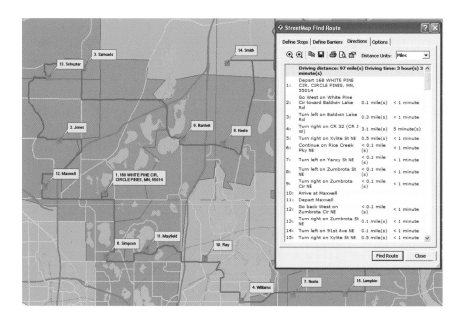

Observe the map and the obvious increased efficiency of the new route. Click the Directions tab in the StreetMap Find Route dialog box and note how much shorter and less time consuming the new route is. This improved efficiency lowers service fleet operational costs and significantly reduces the drive time to service time ratio of each technician. It also enables more service calls per day, resulting in increased customer service and higher levels of customer satisfaction. These differences represent substantial service cost savings to Living in the Green Lane and make a significant contribution to the company's ROI in Business Analyst Desktop and your services.

The Route Finder tool offers significant savings daily to LITGL. When integrated with more sophisticated real-time scheduling systems and GPS-equipped service trucks, these savings become even more significant as dispatchers can match technicians' skills with service call requirements, and adjust routing for real-time weather and traffic patterns.

7. Save the map file as **SalesMgtRoute1.mxd** to C:\My Output Data\Projects\LITGL Minneapolis St Paul\ChapterFiles\Chapter8\.

This completes your work in the suite of Business Analyst Desktop tools designed to support the sales and service management functions. With these solutions in place, you now have used Business Analyst Desktop to complete Living in the Green Lane's comprehensive business model. In the following chapters, you will use additional Business Analyst Desktop tools to identify opportunities for expanding this business model to a national level.

ROI considerations

The integrated business GIS analyses in this chapter support two different revenue streams. The most immediate stream is also the most concrete and easy to calculate. This is the combination of decreased costs and increased revenue attributable to optimizing efficiency in service call routing. Reduced service call costs are the result of lower total operating costs due to shorter driving distances in each day's route for each service vehicle. Conservatively assuming vehicle operating costs of 50 cents per mile, each 20-mile differential achieved by optimized routing reduces costs by $10 per day, or $7,500 per year over LITGL's modest service fleet of three pest-control vehicles. If similar savings are realized in routing the six-person sales force and the three lawn-and-garden maintenance teams, that figure increases to $30,000 for each 20-mile average daily improvement in route efficiency. Recall that the differential in route length you achieved in the analysis above is several multiples of this 20-mile threshold, resulting in a substantial increase in these estimated savings.

In addition to reduced operating costs, the distance differentials also create additional sales and service time, resulting in more revenue from increased sales and service levels. More efficient service also increases customer satisfaction and enhances customer loyalty. These revenues, in turn, impact the Customer Lifetime Value calculations that drive Customer Relationship Management systems.[1] As this model is also relevant for evaluating the ROI impact of effective territory design, we will integrate these factors into that discussion.

Customer Lifetime Value (CLV) models view customers as partners in a long-term relationship rather than just purchasers in a discrete set of transactions. In that context, the value of each customer is a function of a stream of purchases within that relationship compared with the costs of acquiring and retaining that customer. The model considers patterns of brand loyalty, levels of annual purchases, average retention rates of the customer base, changes in pricing over time, and the company's cost of capital discount rates in determining the net present value of a customer's contribution margin over time. These figures then are compared with customer acquisition and retention costs to estimate the lifetime net present value of the customer base.

Routing optimization impacts CLV by decreasing the costs of customer acquisition through reductions in the ratio between travel and selling time for the sales force. It also impacts CLV by reducing the costs of customer retention through reduction of customer maintenance costs and through increases in the daily capacity of each service technician.

Efficient sales territory design also affects the factors contributing to CLV. By creating balanced, serviceable territories, efficient territory systems maximize each representative's sales time, and—by promoting positive morale—each representative's motivation. Similarly, by maintaining target levels of attention to each customer in balanced territories, these systems improve customer satisfaction and retention rates.

When new sales result in a predictable stream of subscription revenue, as they do in the services provided by LITGL, the CLV model is particularly useful. Let us assume that each new customer's annual expenditure on lawn-and-garden maintenance is the $1,200 average for households in the LITGL's market area. Let's also assume that the success rate for new sales calls is 10 percent and that route optimization and efficient territory design allow one additional sales call per day for each sales representative. The annual increase in sales for the six-person sales force in this scenario is $180,000, with virtually no increase in customer acquisition costs beyond sales commission.

Similarly, for every 1,000 customers in the customer base, each 1 percent increase in the retention rate due to improved service efficiency in routing or greater customer satisfaction from more extensive attention from sales and service personnel results in a $12,000 increase in annual revenue, with no increase in customer retention or maintenance costs. Thus, in this context, optimal sales and service routing, and efficient sales territory design, produce benefits approaching $200,000 per year even with the conservative assumptions applied here.

By contrast, the incremental costs of the integrated business GIS tools used to realize these benefits are minimal. Both extensions used in this chapter are integral parts of the Business Analyst Desktop system. The data is either included in the Business Analyst Desktop dataset or can be integrated from internal customer records. The only incremental direct costs, then, are those of the analyst's time and compensation, which are substantially lower than the projected benefits. Thus, in both these areas, the contribution of efficient sales territory design and optimal sales and service routing to CLV and to the company's profit streams is decidedly positive, as is the resulting ROI.

With the implementation of these sales and service operations, you have completed the transformation of Living in the Green Lane from a green-home center to a green-lifestyle center complete with services delivered directly to customers' homes. In doing so, you have enhanced the company's merchandising strategy, expanded its retail presence with two new stores, and extended its reach with a sales and service organization for natural, organic services for lawn-and-garden maintenance services as well as pest control. In short, you have developed a comprehensive business model based on a sustainable, environmentally responsible platform that is ready for growth beyond its original market area. In the next chapters, you will use the Segmentation Module of Business Analyst Desktop to plan that expansion. Before moving to those tasks, however, take a moment to consider what you have accomplished in chapter 8.

Summary of learning
You have expanded your GIS knowledge by learning:

1. The value of integrated business GIS in specifying sales territory design criteria
2. The benefits of balancing territories using demographic and or internal sales
3. The value of routing algorithms in optimizing route efficiency
4. The resource savings and enhanced service capabilities resulting from route optimization

You have enhanced your GIS skills by using Business Analyst Desktop to:

1. Build sales territories around seed points with the Territory Design extension
2. Designate attributes to use in territory balancing schemes

3. Create multilevel sales organizational schemes
4. Realign sales territories to meet organizational objectives
5. Designate stops on a service technician's route
6. Optimize route efficiency with Business Analyst Desktop's Find Route tool
7. Determine efficiencies and cost savings resulting from route optimization

Notes

1. Rust, Roland T., Katherine N. Lemon, and Valarie A Zeithaml. 2000. *Driving Customer Equity: How Customer Lifetime Value is Shaping Corporate Strategy*. New York: Simon and Schuster, pp. 32–45.

Part VI

Customer profiling and segmentation with the ESRI Business Analyst Desktop Segmentation Module

Relevance	Businesses wishing to increase sales to their existing customer base by understanding them more completely, serving them more efficiently, and communicating with them more effectively. Businesses wishing to expand geographically by identifying concentrations of attractive potential customers and estimating sales volume for new stores in their vicinity.
Business scenario	LITGL has completed the transition to a green-lifestyle center business model and expanded its operations to three stores in the Minneapolis-St. Paul area. Management wishes to increase sales by greater sales penetration within existing market areas, as well as geographic expansion to new areas across the United States with attractive population characteristics relative to the company's customer profile.
Analysis required	LITGL must profile its existing customer base in more detail to identify opportunities for greater sales penetration. It also must identify the most profitable segments among its customers and identify geographic concentrations of similar households large enough to support multiple new green-lifestyle centers.
Role of business GIS in analysis	Create profiles of existing customers and base populations using the Segmentation Module and Address Coder. Generate reports detailing the characteristics of the customer base and its most attractive Tapestry Segmentation segments. Use comparative profile reports, maps, and charts to aggregate attractive Tapestry Segmentation segments into Core, Developmental, and Niche target groups. Use Market Potential Indexes to explore the values, lifestyles, purchasing patterns, and media exposure of target groups. Use target group profiles as models, identify geographic concentrations of potential customers, and estimate sales volume for new centers located near them. Determine the feasibility of opening LITGL centers in geographic areas with promising population profiles, Create a segmentation study reporting the findings of these analyses to management and potential investors.
Integrated business GIS Tool	ESRI Business Analyst Desktop, Segmentation Module extension.
ROI considerations: Cost of business GIS	Purchase of Business Analyst Desktop and Segmentation Module, salary of GIS analyst.

ROI considerations: Benefits of business GIS	Increased sales through more effective and efficient merchandising, promotion, and customer service. Increased sales from properly sited new stores near areas of high expected volume.
Environmental impact of business decision	Greater leverage of resources in serving customers of existing facilities. More efficient use of new resources resulting from optimal site selection relative to concentrations of customers.

Table VI.1 Executive summary

The Living in the Green Lane scenario

The Living in the Green Lane green-lifestyle center business model is now fully developed. It provides a comprehensive range of green products and services from modestly sized, strategically located retail centers, each of which exemplifies and demonstrates green-building techniques and living patterns. The company's three stores have been successful and each has a large and growing Green Living Club loyalty program.

Janice and Steven believe that the success of the LITGL concept in Minneapolis-St. Paul underscores the general attractiveness of the business model. They wish to continue the company's growth with a combination of penetration and expansion strategies.

The penetration strategy will focus on increasing sales of existing stores by reaching new customers in their market areas while also expanding sales to existing customers. Implementation of this strategy requires a more complete understanding of existing customers, assessment of the number and sales potential of additional prospects in the stores' market areas, and increased knowledge of the lifestyles, shopping patterns, and media exposure of target customers.

The expansion strategy also will focus on identifying attractive new geographic markets for Living in the Green Lane centers. This involves finding geographic concentrations of potential customers across the United States and assessing their ability to support multiple LITGL centers.

Janice and Steven have determined that service to a new market area will require at least four LITGL green-lifestyle centers given the company's emphasis on high-quality service to a local market. The centers can be company-owned or owned by local partners through franchise agreements. To establish a local presence, they have decided that at least two centers in each market are to be company-owned, though any remaining centers could be either company-owned or franchised.

Based on their experience in the Minneapolis-St. Paul market, Janice and Steven believe that investment in a company-owned center is only feasible if the 3-mile ring around the proposed center has $15 million or more in annual home-improvement materials expenditures.

Franchisees can often use facilities and assets they already own and operate, reducing initial setup costs significantly. Thus, franchised centers are considered viable if home improvement purchase levels are $10 million a year or higher.

Thus, the threshold for the expansion strategy is clear; a market area must be able to support at least two company-owned centers and at least two additional centers, which may be either company owned or franchised. Finding such market areas for new centers is the objective of the expansion component of Janice and Steve's growth strategy.

With these goals in mind, Living in the Green Lane's management team has directed you to use business GIS tools to identify opportunities for sales growth stemming from greater penetration of existing markets and identification of the most attractive opportunities for geographic expansion beyond the Minneapolis-St. Paul area and state. In chapters 9 and 10, you will use the ESRI Segmentation Module of Business Analyst Desktop to accomplish these objectives.

Integrated business GIS tools in customer profiling

You already have used customer profiling and segmentation tools in this book to develop Living in the Green Lane's business model and marketing strategy. The Segmentation Module is an extension to Business Analyst Desktop that significantly extends these capabilities by:

1. Automating the process of customer profiling
2. Supporting multiple sources of data for generating profiles
3. Automating the spatial overlay process for attaching data attributes to customers
4. Providing preformatted reports for describing profiles
5. Providing a systematic, flexible way of forming target groups from Tapestry Segmentation segments based on profile data
6. Incorporating volume data (sales, visits, orders) into segmentation analysis to capture concentrations of activity in Tapestry Segmentation segments
7. Providing penetration and gap data that identify groups of attractive, but underserved, customers
8. Providing reports and mapping functions for estimating sales volume in new geographic markets based on customer profile and volume data
9. Producing customizable, comprehensive segmentation studies that capture the specific data, reports, and charts best communicating the key points of a business GIS research project

In addition, the Segmentation Module provides the highest level of accuracy in profiling and spatial overlay operations as it functions primarily at the block group level. This is the level at which Tapestry Segmentation segments are assigned, so it is the most precise data possible. Furthermore, with the Segmentation Module, Tapestry Segmentation data at the block group level is exposed to all Business Analyst Desktop tools. If the Segmentation Module is not installed, this data is not available below the census tract level.

The Segmentation Module offers several options for customer profiling. One of these is Address Coder. Available as a standalone product, this system also is included in the Segmentation Module. Address Coder performs geocoding, spatial overlay, profiling, and reporting functions through a wizard-style command structure. It begins with a list of customer names and addresses, which can be stored in a variety of data formats, including Microsoft Excel worksheets and Microsoft Access database tables. The system automatically geocodes these addresses and determines the block group in which each address is located. It then provides the option to attach Tapestry Segmentation and demographic attribute fields to each record and to designate a volume field such as sales, orders, or visits. It also offers the option to generate a range of profiling reports and to create a Business Analyst Desktop data layer from the geocoded customer records. Thus, while Address Coder can provide significant customer information directly, it can also be integrated into Business Analyst Desktop for further analysis.

An alternative customer profiling approach is to use Business Analyst Desktop to set up a customer layer from the same type of customer address list used by Address Coder. This process produces a customer layer in the map, which the Segmentation Module then can use to create a profile of existing customers. Although this process does not produce the reports generated by Address Coder, it does provide core data for similar reports produced by other Segmentation Module tools.

Another profiling option is to create profiles of geographic areas such as counties, states, CBSAs, or ZIP Codes. These profiles provide useful comparative possibilities, but also allow users to create customized base profiles to which customer and/or target group profiles can be compared. You will create a profile of the Minneapolis-St. Paul CBSA for precisely this purpose.

Finally, profiles can be created from Mediamark Research Inc. (MRI) survey data in the form of Market Potential Indexes (MPIs), which are reported for each of the 65 Tapestry Segmentation segments. You used these values in your customer profiling tasks in chapter 6. They report the relative frequency with which a segment engages in a certain behavior relative to national averages. In the Segmentation Module application, you could create a profile of people who responded that they "purchased organic soil additives in the past 12 months." Producers of organic soil additives may well be interested in the characteristics of these consumers relative to the general population in which they live. This type of profile facilitates that comparison.

However they are produced, profiles generated by the Segmentation Module provide foundational data for the comparisons that the system generates in report, map, and chart form with its segmentation tools.

Integrated business GIS tools in segmentation analysis

The integrated business GIS tools in the Business Analyst Desktop Segmentation Module extend the segmentation analysis you performed in chapter 6 by integrating sales, demographic, and Tapestry Segmentation data into a cohesive scheme for identifying and serving target groups. In chapter 6, you used a usage segmentation approach to assign customers to segments based on annual purchases, then determined the characteristics of each segment.

The Segmentation Module automates and extends that process while including additional data sources in the analysis. Its central function is to compare profiles of existing customers with a relevant population base, in this case the Minneapolis-St. Paul CBSA, to determine the distinguishing characteristics of the customer base.

The primary basis for comparison is Tapestry Segmentation composition. By using this system as its foundation, the Segmentation Module brings to the process not only extensive descriptions of the characteristics, behaviors, and lifestyles of these segments, but also the Market Potential Indexes that provide detailed data on purchase and media exposure patterns. This integration increases the depth of insight into current customers and their behavior. It also expands the horizon of the segmentation process from a descriptive function to a prescriptive one. That is, it provides analyses that identify growth opportunities for the company and project their impact.

You will use these capabilities to explore two growth strategies in chapter 10. The first is a penetration strategy with the objective of increasing Living in the Green Lane's sales within the market areas of its three existing stores. The customer insight and information generated by the charts, maps, and reports will guide this effort. The information these documents provide on customer location, concentration, behavior, purchasing, and media exposure patterns will allow you to refine LITGL's marketing and promotion strategies to serve their customers more effectively. This will increase both the number of customers in current market areas and their average purchases.

The second growth strategy is geographic expansion with the objective of finding the best geographic regions in the United States for Living in the Green Lane's planned new facilities. Using the Market Potential Volume Report, you will project the level of LITGL's expected sales in each of the 940 CBSAs in the United States.

This data will help you select the most favorable CBSA for the initial development of new LITGL green-lifestyle centers. In addition to the report, the Segmentation Module produces map layers of relevant data for display in Business Analyst Desktop. This resource allows you to drill down more deeply to assess the specific capacity of individual CBSAs.

By focusing on a specific CBSA and repeating the Market Potential Volume Report at the census tract level, you will generate a map layer displaying the distribution of expected sales by census tract within the CBSA. This will help you determine whether demand is sufficiently dispersed within the market area to support the objective of four new LITGL stores. You also will be able to assess which market areas within the CBSA can support company-owned stores and which are better served with franchise agreements with local partners.

In short, using the combination of profiling and segmentation tools within the Segmentation Module, you will be able to increase Living in the Green Lane's knowledge of its customer base and use that knowledge to guide the company's growth strategies.

Chapter 9

Creating customer profiles

The Living in the Green Lane business model is now complete. The company has grown to three locations, while enlarging and fine tuning its product line. Janice and Steven are now ready to expand into new markets around the United States. To do so, they have directed you to create a profile of your best customers, identify the most profitable segments, and seek out concentrations of similar prospective customers in other parts of the country. You will use the Business Analyst Desktop Segmentation Module to do so.

The first step in this process is to create a customer profile from internal records of your best customers. These are the members of the Green Living Club, which now has members affiliated with each of the three stores. You will compare this profile to the general population characteristics of the Minneapolis-St. Paul area to identify segments particularly attracted to LITGL. You will use this information, along with sales data, to select the Tapestry Segmentation segments that will constitute the target market for national expansion.

Creating and comparing profiles of customer groups, geographic areas, and buyer-survey respondents also provides useful insight into buying characteristics. You will create and compare several profiles to illustrate this application.

Run Business Analyst Desktop; load map

1. Click Start, Programs, ArcGIS, Business Analyst, BusinessAnalyst.mxd to run ArcMap, load the Business Analyst Extension, and then load the default Business Analyst map.

2. Click OK when the Update Spatial Reference dialog box appears, then close the Business Analyst Assistant window on the right of the screen.

3. Click File, click Open. Navigate to C:\My Output Data\Projects\LITGL Minneapolis St Paul\CustomData\ChapterFiles\Chapter9\LITGLProfile.mxd. Click the map file to open it.

This map contains a layer displaying available Living in the Green Lane's Steiers, Mason, and Longwell stores over a thematic map of Home Related Expenditures per Household by block group. The data source for this layer is a shapefile similar to the one you created in chapter 3. You will use this map to display the members of the Green Living Club for each store and determine their distinguishing characteristics.

Create profile for the Minneapolis-St. Paul CBSA

The Segmentation Module provides several methods for creating population profiles. The first method you will use calculates profiles for geographic areas. This method serves two purposes: First, it can be used to create profiles for trade areas in the absence of customer data. In this application, the result would be a profile of all the households within, for example, a five-minute drive time of a potential site.

Second, this method can create base profiles of geographic areas that can be compared with customer profiles. This allows the analyst to identify distinguishing customer characteristics relative to a comparable general population. In this instance, you will create a base profile of the Minneapolis-St. Paul CBSA with which to compare the profile of Green Living Club members.

1. Click the drop-down menu on the Business Analyst toolbar, click Segmentation, then click Create Profiles to open the Segmentation Wizard. Select the Create a geographic profile option, click Next.

9

10

2. In the next window, select LITGLMinnStPaul as the boundary layer to profile and Total Households as the profile base. Recall that in chapter 2 you defined this study area layer as the Minneapolis-St. Paul CBSA. Households is the appropriate profile base as you will be using Tapestry Segmentation household data throughout the profiling and segmentation application. When the window resembles this, click Next. If you receive as Spatial Index notice, click Yes.

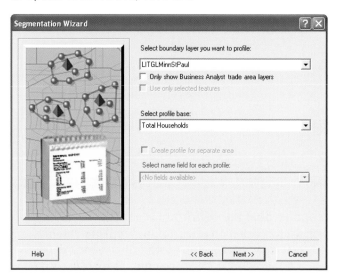

3. In the next window, enter **MinnStPaulCBSA** as the name of the new Segmentation Profile. Click Finish.

The Segmentation Module calculates the customer profile and displays it in a window similar to the one below. This is an XML file, which can also be displayed in Microsoft Excel if you wish.

Segment ID	Segment Name	Count	Percent	Total Volume	Average Volumetric	
1	Top Rung	5300	0.413	5301	1	
2	Suburban Splendor	50353	3.919	50353.02	1	
3	Connoisseurs	10652	0.829	10653	1	
4	Boomburbs	93478	7.275	93478	1	
5	Wealthy Seaboard Suburbs	7612	0.593	7613	1	
6	Sophisticated Squires	147400	11.472	147400.05	1	
7	Exurbanites	41105	3.199	41105	1	
8	Laptops and Lattes	4711	0.367	4711	1	
9	Urban Chic	8559	0.666	8559	1	
10	Pleasant-Ville	3786	0.295	3787	1	
11	Pacific Heights	0	0	0	0	
12	Up and Coming Families	103287	8.039	103287.99	1	
13	In Style	74032	5.762	74032	1	
14	Prosperous Empty Nesters	19846	1.545	19847	1	
15	Silver and Gold	1052	0.082	1053	1	
16	Enterprising Professionals	45851	3.569	45851.02	1	
17	Green Acres	50034	3.894	50035	1	
18	Cozy and Comfortable	74048	5.763	74048.01	1	
19	Milk and Cookies	16265	1.266	16265	1	
20	City Lights	0	0	0	0	
21	Urban Villages	0	0	0	0	
22	Metropolitans	51022	3.971	51022.99	1	

Review the contents of this window. Each of the 65 Tapestry Segmentation segments is listed, as well as the number of households in the CBSA within each segment. The Percent column reports this number as a percentage of all households in the CBSA. As this is not a customer profile, there is no volume measure such as sales, orders, or store visits included. Thus, the Total Volume column simply repeats the Count column. This profile is now ready to be used as a base profile in subsequent segmentation analysis.

Create a customer profile with Address Coder

Address Coder is a standalone application that creates profiles from tables of customer data. Its functionality has been integrated into the Business Analyst Desktop Segmentation Module and you will use it to create a profile from a table of Green Living Club members and generate a series of reports on them.

1. Close the CBSA Profile in the Segmentation window. Click the drop-down menu on the Business Analyst toolbar, click Segmentation, then click Create Profiles to open the Segmentation Wizard. Select the Create a profile using Address Coder, then click Finish.

2. The window closes and the main Address Coder wizard opens. Click the Browse button to the right of the Input box, then navigate to C:\My Output Data\Projects\LITGL Minneapolis St Paul\Custom Data\ChapterFiles\Chapter9\. In the file type drop-down menu at the bottom right of the window, select Microsoft Excel 97-2003 (*.xls) as the file type. The file LITGLCustomersFull.xls will appear in the window. Select this file then click Open to select this as the input file. In the Select Sheet window, select the 'LITGL Customers$' option, then click OK. This file contains a list of more than 1,800 Green Living Club members, including some from each of the three stores. As this is a hypothetical list, it contains no names or ZIP Codes.

3. The Output destination folder is fixed. The Report file should be saved to the C:\My Output Data\Projects\LITGL Minneapolis St Paul\CustomData\ChapterFiles\Chapter9\ folder. To specify this location and filename, click the Browse button to the right of this box, navigate to this folder and enter **LITGLCustomerProfileAC.doc** as the Report destination file. Your window should resemble this one. When it does, click Next to advance from the Files tab to the Fields tab.

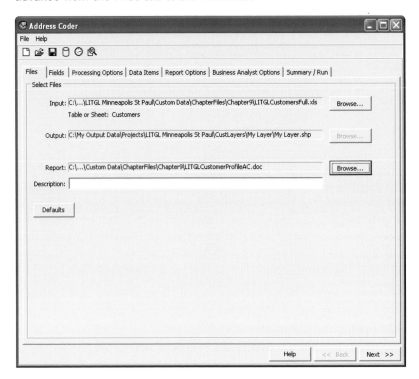

4. The Fields tab designates the fields in the MS Excel file to be used for geocoding the addresses. Accept the defaults, then click Next to move from the Fields tab to the Processing Options tab. Accept the defaults in this tab as well, and click Next to move to the Data Items tab.

5. The Data Items tab allows you to select attributes to be appended to each customer record in the output data file. Expand the list of attributes in the left window and move the following attributes to the right by selecting them and clicking the single right arrow key.

    Tapestry Code
    Tapestry Life Mode Code
    Tapestry Urbanization Code
    CY Average Household Size
    CY Median Age
    CY Education Index

CY Median Net Worth
CY Median Household Income
CY Median Value of Owner Occupied Hus

When the window resembles the one below, click Next to move to the Report Options tab.

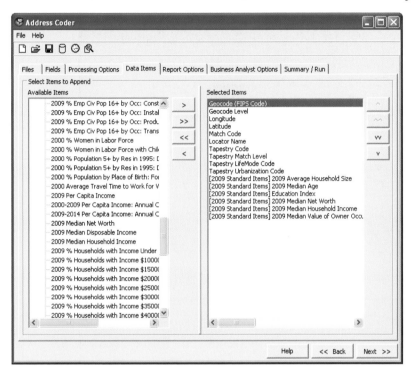

6. In the Report Options tab, select all the available reports. Make sure Sort on: is set to Numerical Segment Order Asc and Life Mode is the Summary Group to be used.

    Do not select the Create Tapestry CSV File option, the Distribute uncoded records across profile, or the Show Composition of Custom Base options. Click Next to move to the Business Analyst Options tab.

7. In the Business Analyst Options window, select Cust ID as the Customer Name field, Store as the Store ID field, and LITGL Minneapolis-St. Paul as the Business Analyst Project. Enter **LITGLCustomerLayer** as the Layer Name, but do not select the Create Business Analyst Tapestry Profile File option.

When the BA tab resembles the one below, click Next.

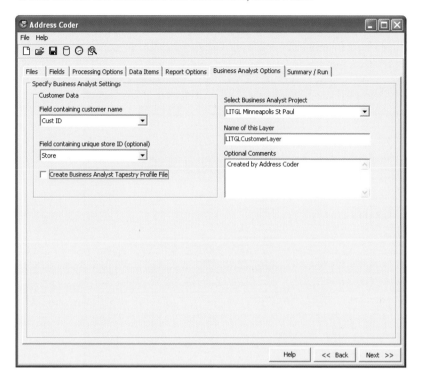

8. A Summary window that displays the settings appears. To run the analysis with these settings, click Run.

The window displays processing progress until a window opens that asks for the Base you wish to use for the Customer Tapestry Profile. Select the United States option. (This will create comparisons between LITGL's customer base and the population of the United States. In later reports, you will make similar comparisons with the population of the Minneapolis-St. Paul CBSA. The latter is more useful in understanding local customer characteristics and the former for identifying expansion opportunities at the national level. The last option creates a profile based on only those block groups in which at least one customer resides, the most limited geographic area available.) Click OK to complete the analysis.

The Segmentation Module completes the analysis, opens a Microsoft Word window displaying the reports you selected, and displays a message indicating that Address Coder will close—unless you specify otherwise—after processing in order to return to Business Analyst. Accept the default by clicking OK. Address Coder closes and returns you to Business Analyst, where the customer layer is displayed on the map. Select the Microsoft Word window and scroll through the Customer Profile Reports. You will find a comprehensive set of demographic data on the customer base and a series of reports on the geographic distribution of customers.

Toward the end of the list you will find two reports to consider carefully. The first is the Customer Tapestry Profile report, which should resemble the one below.

ESRI

Customer Tapestry Profile
Page 2 of 4

File: C:\My Output Data\Projects\LITGL Minneapolis St Paul\CustLayers\LITGLCustomerLayer\LITGLCustomerLayer.shp

Number of Records: 1,864

Tapestry Description	Customers Number	%	Penetration Per 100	U.S. Number	%	Index
1 Top Rung	7	0.4	0.24	2,866	0.2	250
2 Suburban Splendor	228	12.4	1.28	17,831	0.9	1307
3 Connoisseurs	19	1.0	0.26	7,339	0.4	265
4 Boomburbs	280	15.2	1.20	23,415	1.2	1222
5 Wealthy Seaboard Suburbs	3	0.2	0.00	0	0.0	0
6 Sophisticated Squires	281	15.3	1.61	17,431	0.9	1647
7 Exurbanites	235	12.8	0.72	32,536	1.7	738
8 Laptops and Lattes	0	0.0	0.00	1,752	0.1	0
9 Urban Chic	43	2.3	1.31	3,277	0.2	1341
10 Pleasant-ville	1	0.1	0.00	0	0.0	0
11 Pacific Heights	0	0.0	0.00	0	0.0	0
12 Up and Coming Families	107	5.8	0.20	53,816	2.9	203
13 In Style	248	13.5	0.71	34,740	1.8	729
14 Prosperous Empty Nesters	11	0.6	0.03	37,994	2.0	30
15 Silver and Gold	1	0.1	0.03	3,369	0.2	30
16 Enterprising Professionals	96	5.2	0.51	18,665	1.0	526
17 Green Acres	11	0.6	0.02	62,901	3.3	18
18 Cozy and Comfortable	59	3.2	0.23	25,606	1.4	235
19 Milk and Cookies	45	2.4	0.17	26,697	1.4	172
20 City Lights	0	0.0	0.00	0	0.0	0

This report lists the number of customer households in each Tapestry Segmentation segment and the percentage of all customers in that segment. The Number and Percent (%) columns to the right display the same values for the population of the United States. The Index column reports the relative presence of a segment within the customer base compared with that segment's presence across the United States. Numbers greater than 100 indicate a segment is more prevalent within the customer base than within the nation. Index values below 100 indicate the reverse. Thus, the segments with the highest index values above 100 represent concentrations of segments in the customer base compared with the general population.

9

10

The second report to review is the Top 20 Tapestry Segments by Customer Count portion of the Customer Tapestry Profile report. It should resemble the one below.

ESRI

Customer Tapestry Profile
Top 20 Tapestry Segments by Customer Count
Page 4 of 4

File: C:\My Output Data\Projects\LITGL Minneapolis St Paul\CustLayers\LITGLCustomerLayer\LITGLCustomerLayer.shp

Number of Records: 1,864

Rank	Tapestry Description	Customers	U.S.	Index
1	6. Sophisticated Squires	15.3%	0.9%	1647
2	4. Boomburbs	15.2%	1.2%	1222
3	13. In Style	13.5%	1.8%	729
4	7. Exurbanites	12.8%	1.7%	738
5	2. Suburban Splendor	12.4%	0.9%	1307
	Subtotal	69.1%	6.7%	1032
6	12. Up and Coming Families	5.8%	2.9%	203
7	16. Enterprising Professionals	5.2%	1.0%	526
8	18. Cozy and Comfortable	3.2%	1.4%	235
9	24. Main Street, USA	2.6%	0.0%	0
10	19. Milk and Cookies	2.4%	1.4%	172
	Subtotal	19.3%	6.6%	291
11	9. Urban Chic	2.3%	0.2%	1341
12	30. Retirement Communities	2.1%	0.4%	578
13	3. Connoisseurs	1.0%	0.4%	265
14	28. Aspiring Young Families	1.0%	1.2%	80
15	36. Old and Newcomers	0.7%	1.0%	69
	Subtotal	7.2%	3.2%	226

This report identifies the top 20 segments in LITGL's customer base ranked by number of customers. The percentage of each segment within the customer base is reported, as is the comparable number for the United States and the Index representing the relationship between the two. All segments in blue are more prevalent within the customer base than in the national population, hence the Index values greater than 100.

This report also indicates the concentration of the LITGL's customers by Tapestry Segmentation segment. Note that the top five segments account for almost 70 percent of customers and the top 10 almost 90 percent. This means two things:

First, Living in the Green Lane can concentrate its marketing efforts on a relatively small number of segments that represent a relatively large number of its customers. Second, these segments are good candidates for selection as target groups for future stores. To make that determination, their purchase behavior must be considered as well as their numeric size. The Segmentation Module provides other tools to include this crucial factor in the analysis.

Create a profile from a customer layer

The Segmentation Module provides the ability to generate a customer profile from a customer data layer. The LITGLCustomer Layer you created with Address Coder provides such a profile. However, you wish to add a volumetric component to that profile in order to include sales information in the analysis.

1. Click the drop-down menu on the Business Analyst toolbar, click Segmentation, then click Create Profiles to open the Segmentation Wizard. Select the Create a profile using customer data option, then click Next.

2. In the next window, select the Use an existing customer layer option, then use the drop-down menu to select the LITGLCustomerLayer layer. If you receive a spatial index message, click Yes, then click Next. In the next window, select the Use volume information option, use the drop-down menu to select LYPurchase as the volume field, and select Total Households as the Profile base. When the window resembles the one below, click Next.

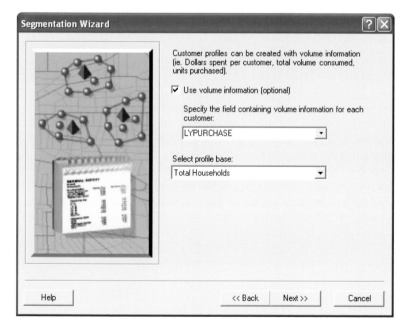

3. In the next window, enter **LITGL Customer Profile** as the name of the Segmentation Profile, then click Finish.

The Segmentation Module generates the profile using the data in the customer layer and opens a window displaying the contents of the xml file that contains the profile. This is similar to the one you produced before with Address Coder. Note, however, the additional attributes that report total and average household purchases from Living in the Green Lane for each customer segment.

4. Click twice on the Total Volume column header to sort this field in descending order. The table should now resemble the one below, with data sorted by total volume. The Average Volumetric attribute reports average purchases at LITGL per household in each segment.

Segment ID	Segment Name	Count	Percent	Total Volume	Average Volumetric
4	Boomburbs	285	16.447	3103229	10888.52
2	Suburban Splendor	230	16.142	3045679.06	13242.08
6	Sophisticated Squires	282	16.005	3019969.98	10709.11
13	In Style	252	13.029	2458372.23	9755.45
7	Exurbanites	236	12.978	2448737.26	10376.01
12	Up and Coming Families	110	6.389	1205516.23	10959.24
16	Enterprising Professionals	94	4.293	809957.55	8616.57
18	Cozy and Comfortable	59	2.381	449316	7615.53
30	Retirement Communities	40	2.239	422473.3	10561.83
9	Urban Chic	43	2.222	419257.26	9750.17
24	Main Street, USA	49	2.001	377520.32	7704.5
19	Milk and Cookies	44	1.376	259553.94	5898.95
28	Aspiring Young Families	18	0.837	157945.93	8774.77
14	Prosperous Empty Nesters	11	0.546	103034.95	9366.81
39	Young and Restless	13	0.479	90470	6959.23
36	Old and Newcomers	14	0.423	79721	5694.36
3	Connoisseurs	19	0.367	69218	3643.05
17	Green Acres	11	0.272	51321	4665.55
65	Social Security Set	7	0.238	44954	6422
29	Rustbelt Retirees	4	0.199	37616	9404
5	Wealthy Seaboard Suburbs	3	0.181	34223.67	11407.89
1	Top Rung	7	0.169	31823	4546.14

You wish to include among LITGL's target segments those that constitute at least 4 percent of the company's customer base and have high average annual purchasing levels. Use the table below to record the segment ID, percent, and average volumetric values for segments, which account for 4 percent or more of LITGL's customer base. You will use this table when selecting the company's target segments, a task to which you will now turn. When you have completed the chart, close the profile window.

Segment ID	Percent	Average Volumetric

Table 9.1 Segments accounting for 4 percent or more of LITGL's customer base

Create target group from profile information

You have used several alternative techniques to develop profile information for use in your segmentation analysis. Specifically, you have created profiles for Living in the Green Lane's loyalty club customers and the Minneapolis-St. Paul CBSA. You will now use this information to select LITGL's target segments for future marketing and expansion initiatives.

1. Click the drop-down menu on the Business Analyst toolbar, click Segmentation, then click Create Target Groups to open the Segmentation Wizard. Select the Create target group from a chart option, then click Next.

2. In the resulting window, select the Create target group using existing profiles to create a new chart option. Click the Add button and add LITGL Customer Profile to the profile box. Select MinnStPaulCBSA in the base profile box and Lifemodes as the target group. When your window resembles the one below, click Next.

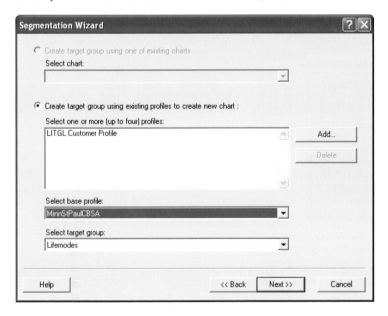

In the resulting window you will name your target group and select the Tapestry Segmentation segments that will comprise it. You will do so using two graphic aids and the table you completed on page 226.

3. Drag and drop the corners of this window to increase its visibility. The window displays a bar chart of Tapestry Segments. The meaning of the chart is discussed below, but you wish to set the width of each bar proportional to the segment's composition of LITGL's customer base. To do so, click Properties, then click the Scaling tab. In the top part of the dialog box, click the Percent Composition option. In the Chart Axes box at the bottom of the window, select the "Index" & "% Composition Volume" option, a setting that will affect the Game Plan Chart, then click OK.

The chart now resembles the image below, though the colors may differ. Move the cursor over one of the bars in the chart to view the callout box, which displays data for that bar as illustrated below.

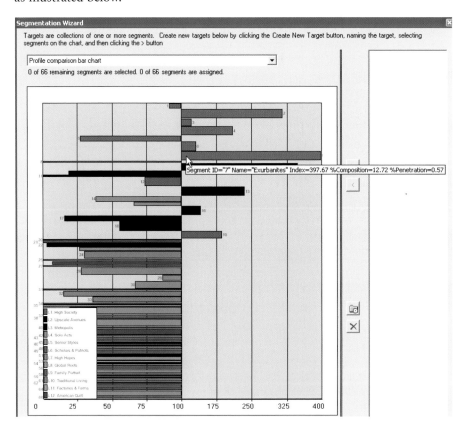

Each bar in this chart represents a Tapestry Segmentation segment. The color of the bar indicates the Life Mode group to which it is assigned. The length and direction of each bar reflect the index of the Tapestry Segmentation segment in LITGL's customer base. The central point is 100. Bars extending to the right represent segments that are more prevalent in LITGL's customer base than they are in the general CBSA population. Bars extending to the left indicate the reverse. The width of each bar represents the percentage of LITGL's customer base within this segment. Thus, longer and broader bars extending to the right are the most attractive segments numerically.

You could add segments to the target group from this window, but you will view another chart before doing so.

4. Click the drop-down arrow in the field at the top of the window and select Game Plan Chart as the chart type. The chart will change to one that resembles the image on page 229. In this chart, segments are represented as circles whose colors represent the Life Mode groups to which they belong. Move the cursor over a segment circle

to see the callout box that displays segment data as illustrated below. To make the placement of segments exactly match the figure, change the x-axis to 4 and the y-axis to 110 in the boxes at the bottom of the chart.

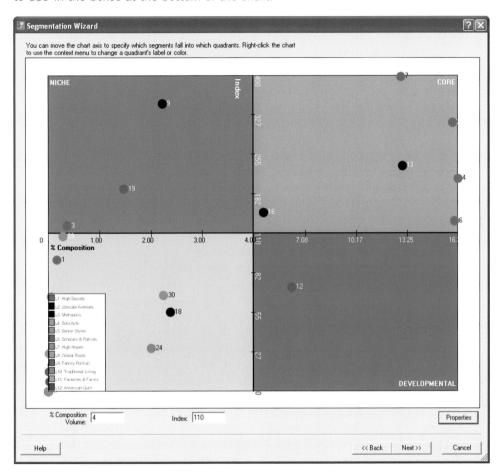

This chart is divided into four quadrants, each representing segments with different characteristics. The chart is defined by two values: Percent (%) Composition Volume and Index. Percent Composition Volume is the volume of each segment's purchases as a percentage of LITGL's customers' total purchases. The x-axis value of 4, which you can adjust if you wish, indicates that segments representing 4 percent or more of the total sales are in the right two quadrants, while those below that are in the left two quadrants.

The second defining value is Index. Recall from the reports you created earlier that Index compares a segment's percentage of LITGL's customer base with its percentage of the general population in the base profile area, in this case the Minneapolis-St. Paul CBSA. Segments with values over 100 are more prevalent in the customer base than in the general population. The reverse is true for segments with Index values below 100. The y-axis value of 110, which you can also adjust, indicates that segments with Index values above 110 are in the top two quadrants, while those with Index values below 110 are in the bottom two.

Clearly, then, by changing the values in the x-axis and y-axis settings, you can change the values that define each of the four quadrants.

The Game Plan Chart uses these values to define three groups of customer segments:

Core segments are those with values above the boundary in both measures. Thus, they represent a substantial percentage of LITGL's total sales and are more prevalent there than in the general CBSA population.

Developmental segments are those that represent a substantial portion of total sales but are less prevalent in that customer base than in the general CBSA population. Although significant in number, they are not as attracted to LITGL's stores as are Core segments.

Niche segments are more prevalent in LITGL's customer base than in the general CBSA population. However, they do not comprise a substantial percentage of the total sales. Niche segments might present attractive opportunities if they are present in a market area, but should not have the priority of Core customer segments.

Take a moment to identify the Core and Developmental segments in the chart. Review the average purchases per household data in the table you created above to determine the purchasing levels of each of these segments. As all the segments in these categories have high purchasing levels, you will include them in the target group. You also wish to include the Niche segment with the highest index value in the target group. Based on the chart, that is segment 9: Urban Chic. Using this chart, you will create a target group that includes these segments.

5.   Click the Create new target button [icon] at the right of the chart and, at the top of the box on the right, enter LITGL **Target Segments** as the name of the group. Press the Ctrl key, which enables the selection of multiple symbols, and click the dots that are in the Core and Development segments as well as segment 9 in the Niche quadrant to select them. Click the right arrow key to move these segments to the right box where they will appear under the new group name. When the window resembles the one on page 231, click Next.

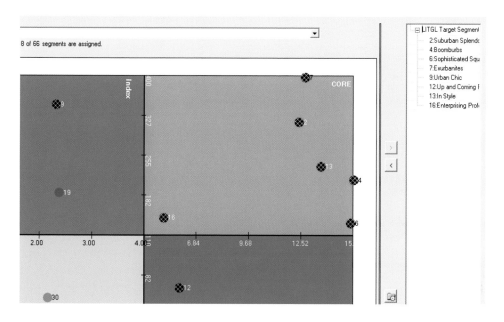

6. In the resulting window enter **LITGL Target Segments** as the name of the new target group, then click Finish.

The Segmentation Module creates the new target group and displays a window to indicate that this procedure was successful. Though not displayed on the map, the Target Group definition is ready for use in further segmentation analyses. Close the window and save your work.

You have used Address Coder and the Business Analyst Desktop Segmentation Module to create a profile of Living in the Green Lane's most valuable customers—the members of its Green Living Club. You also have used the Segmentation Module to create a base profile of the Minneapolis-St. Paul CBSA for comparative purposes.

Based on this comparison, along with LITGL sales data, you have identified the most attractive Tapestry Segmentation segments in the customer base and designated them as the target group for further segmentation analysis. You will perform that analysis in chapter 10.

Chapter 10

Segmentation analysis for enterprise expansion

While the profiles and target groups you created in chapter 9 provide useful information on Living in the Green Lane's customers, they are also valuable as the data required in segmentation analysis with the Business Analyst Desktop Segmentation Module. In this chapter, you will use the module's tools to design a series of reports, maps, and charts to extend that analysis further. Specifically, you will use these resources to describe your customers more fully, to study their buying behavior in order to serve them more effectively, and to identify opportunities for expansion in other markets nationwide.

These tasks support significant enterprise objectives. Enhanced descriptions of current customers allow Living in the Green Lane to perform more precise targeting. Deeper understanding of buying behavior enables greater penetration of existing markets. Identification of market areas that match customer profiles unveils attractive growth opportunities.

The Segmentation Module facilitates these analyses by comparing the profiles you created in chapter 9 with each other and with units of geography in the current market area and beyond. The table below categorizes reports, charts, and maps in the Segmentation Module by the type of analysis they support. You will create the items marked with asterisks as you work through this chapter. The remaining items will be discussed briefly in each section and their contents summarized.

Targeting	Penetration	Growth
Reports	**Reports**	**Reports**
1. Profile Segmentation 2. Core and Development Segments 3. Geographic Customer Summary 4. Customer Demographic Profile 5. Match Level Summary 6. Profile Volume Segmentation*	1. Market Area and Gap Analysis* 2. Understanding Your Target Customers* 3. Developing Marketing Strategies*	1. Market Potential 2. Market Potential Volume*
		Maps
Charts		1. Target Map 2. Target Penetration Map 3. Four Quadrant (Game Plan) Map*
1. Profile Comparison Bar Chart* 2. Four Quadrant (Game Plan) Chart*		

Table 10.1 Segmentation Module tools for targeting, penetration, and growth

Run Business Analyst Desktop; load map

1. Click Start, Programs, ArcGIS, Business Analyst, BusinessAnalyst.mxd to run ArcMap, load the Business Analyst Extension, and then load the default Business Analyst map.

2. Click OK when the Update Spatial Reference dialog box appears, then close the
    Business Analyst Assistant window on the right of the screen.

3. Click File, click Open. Navigate to C:\My Output Data\Projects\LITGL Minneapolis St
    Paul\CustomData\ChapterFiles\Chapter10\LITGLSegmentation.mxd. Click the map file
    to open it.

This map is similar to the basemap for chapter 9. It contains a layer displaying Living in the
Green Lane's Steiers, Mason, and Longwell stores over a thematic map of Educational Attainment by block group. The data source for this layer is a customized shapefile similar to
the one you created in chapter 3. The map also displays LITGL's customers, color coded by
store affiliation. You will use this map to display the results of Segmentation Module map
analyses.

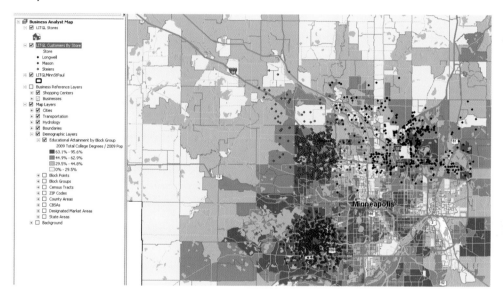

Create segmentation reports and charts for customer targeting

The first major objective of segmentation analysis, then, is to use the descriptive tools of the
Segmentation Module to deepen your understanding of LITGL's customers.

1. Click the drop-down menu on the Business Analyst toolbar, click Segmentation, then
    click Segmentation Charts, Maps, Reports. Click Create new segmentation charts,
    maps and reports, then click Next. Click Segmentation reports, then click Next to open
    the Segmentation Reports dialog box, which lists all the segmentation reports available
    in the module.

2. Select the Profile Volume Segmentation Report, then click Next. In the resulting
    window, Select LITGL Customer profile as the Target segmentation profile and
    MinnStPaulCBSA as the Base segmentation profile. When the window resembles the
    one on page 236, click Next.

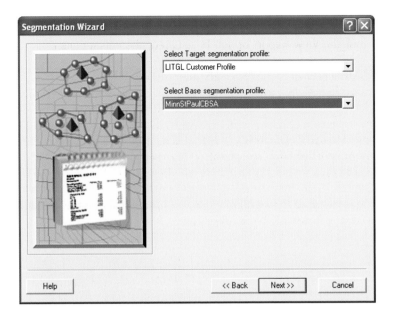

3.   In the next window, enter LITGL Customer Profile as the Target Profile Description, MinnStPaulCBSA as Base Profile Description. Select Total Volume as the sorting attribute and select Descending order. Enter dollars spent as the description of volumetric information. With these settings, the report will compare LITGL's customer profile with the Minneapolis-St. Paul CBSA. It will report the total and average dollars spent by each segment and list segments in order of highest to lowest total purchases. When the window resembles this, click Next.

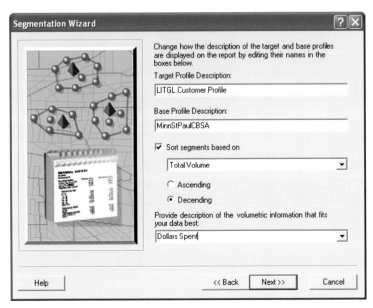

4. In the next window, enter **LITGL Profile Volume Segmentation Report** as the report title and name. Confirm that the View Report option is selected, then click Finish.

   The Segmentation Module performs the necessary analysis and calculations, then displays the report in a window on your screen. Review the report and scroll through its pages by clicking the arrow buttons at the left of the command bar at the top of the screen. The first page explains the contents of the report. The next two pages display results by Life Mode and Urbanization group. The letters in the callout boxes attached to each column are keyed to the explanations in the leftmost column, which are abbreviated versions of the more extensive explanations on the first pages of the report.

   Scroll to the fourth page of the report, which displays values for each of the segments in the customer profile. It should resemble the following.

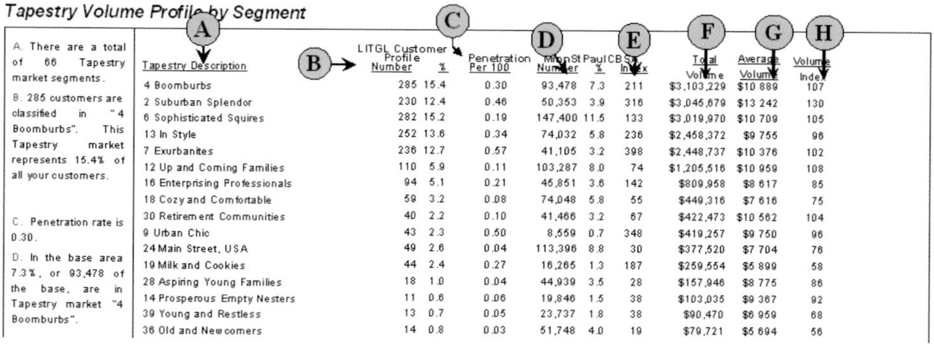

Tapestry Volume Profile by Segment

A. There are a total of 66 Tapestry market segments.

B. 285 customers are classified in "4 Boomburbs". This Tapestry market represents 15.4% of all your customers.

C. Penetration rate is 0.30.

D. In the base area 7.3%, or 93,478 of the base, are in Tapestry market "4 Boomburbs".

Tapestry Description (A)	LITGL Customer Profile Number (B)	LITGL Customer Profile %	Penetration Per 100 (C)	MinnStPaul CBSA Number (D)	MinnStPaul CBSA %	Index (E)	Total Volume (F)	Average Volume (G)	Volume Index (H)
4 Boomburbs	285	15.4	0.30	93,478	7.3	211	$3,103,229	$10 889	107
2 Suburban Splendor	230	12.4	0.46	50,353	3.9	316	$3,045,679	$13 242	130
6 Sophisticated Squires	282	15.2	0.19	147,400	11.5	133	$3,019,970	$10 709	105
13 In Style	252	13.6	0.34	74,032	5.8	236	$2,458,372	$9 755	96
7 Exurbanites	236	12.7	0.57	41,105	3.2	398	$2,448,737	$10 376	102
12 Up and Coming Families	110	5.9	0.11	103,287	8.0	74	$1,205,516	$10 959	108
16 Enterprising Professionals	94	5.1	0.21	45,851	3.6	142	$809,958	$8 617	85
18 Cozy and Comfortable	59	3.2	0.08	74,048	5.8	55	$449,316	$7 616	75
30 Retirement Communities	40	2.2	0.10	41,466	3.2	67	$422,473	$10 562	104
9 Urban Chic	43	2.3	0.50	8,559	0.7	348	$419,257	$9 750	96
24 Main Street, USA	49	2.6	0.04	113,396	8.8	30	$377,520	$7 704	76
19 Milk and Cookies	44	2.4	0.27	16,265	1.3	187	$259,554	$5 899	58
28 Aspiring Young Families	18	1.0	0.04	44,939	3.5	28	$157,946	$8 775	86
14 Prosperous Empty Nesters	11	0.6	0.06	19,846	1.5	38	$103,035	$9 367	92
39 Young and Restless	13	0.7	0.05	23,737	1.8	38	$90,470	$6 959	68
36 Old and Newcomers	14	0.8	0.03	51,748	4.0	19	$79,721	$5 694	56

   Review the explanations of each column to understand the information displayed here. The segments at the top of the page have the highest total purchases of LITGL's products and services. Among these segments, those with Index values greater than 100 comprise a higher percentage of LITGL customers than they do the general CBSA population. Low values in the Penetration-per-100 column indicate significant remaining opportunity to reach prospects in this segment of the CBSA area.

   Note, as well, the Average Volume and Volume Index figures, in the rightmost columns of the report. Average Volume reports the average level of purchases of LITGL products and services by households in each segment. Volume Index values greater than 100 indicate segments with higher average purchase levels than the customer base as a whole. Thus, segments with high Total Volume, Index, and Volume Index values represent a high proportion of LITGL's customer base, account for a significant portion of total sales, and have above average levels of purchases per household. Also note that these are largely the segments you selected as LITGL's target segments in chapter 9.

   If you wish you may print or export this report, though it will also be available to you in the Segmentation Module data files.

5. Close the report window.

Select several other available reports from the list below, using the designated settings to view the other Segmentation Module resources that support the Targeting objective.

6. Go back to the Business Analyst toolbar. Click Segmentation, then Segmentation Charts, Maps, Reports. Select Create new segmentation charts, maps, and reports. Click Next. Click Segmentation Reports, then click Next.

Run **Profile Segmentation Report** with LITGL's Customer Profile and MinnStPaul-CBSA to view the comparisons of the Profile Volume Segmentation Report but without the volume data. This report is useful in segmentation studies lacking purchasing data.

Run **Core and Development Segments Report** with LITGL's Customer Profile and MinnStPaulCBSA. Use the Threshold approach at the default levels to the segments assigned to the Core and Development groups discussed in chapter 9. Compare these segments to those at the top of the Profile Volume Segmentation Report to assess the importance of these groups to Living in the Green Lane's sales performance.

Run **Geographic Customer Summary Report** with LITGL Customers by Store as the Customer layer to generate a report listing the geographic areas in which the customer base is concentrated, including the top 20 counties, ZIP Codes, and CBSAs.

Run **Customer Demographic Profile Report** with Block Groups as the Geography level and LITGL Customers by Store as the customer layer to generate a report summarizing the demographic characteristics of the customer base. This report uses a spatial overlay procedure to attach demographic data to customer records, then summarizes the results to produce the profile report. It is an automated version of the analysis you performed in chapter 6.

Run **Match Level Summary Report** with LITGL Customers by Store as the customer layer and Customer as the customer description to generate a report summarizing the level at which customer addresses were geocoded when the LITGL Customers by Store layer was created. Matches are listed in descending order of accuracy, so that the greater the number of addresses geocoded as Address Points, and the fewer geocoded as ZIP Code Centroid or No Match, the more accurate the estimation of customer location and, therefore, the spatial overlay procedure.

In addition to these reports, the Segmentation Module provides the Profile Comparison Bar Chart to explore customer characteristics graphically.

7. Click the drop-down menu on the Business Analyst toolbar, click Segmentation, then click Segmentation Charts, Maps, Reports. Click Create new segmentation charts, maps and reports, then click Next. Click Segmentation charts, then click Next to open the Segmentation Charts dialog box, which lists the charts available in the module.

8. Select the Profile comparison bar chart, then click Next. In the next window, select LITGL Customer Profile as the Select profile, MinnStPaulCBSA as the base profile, and LITGL Target Segments as the Target group. When the window resembles the one below, click Next.

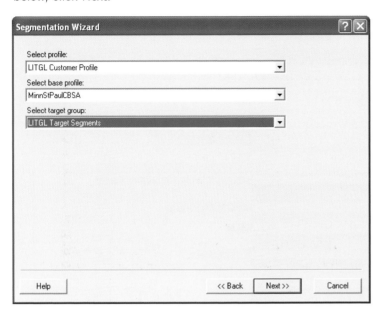

The Segmentation Module applies your settings and displays a bar graph of the Index values of each of the 65 segments for LITGL's customer base. The segments you designated as target segments are displayed in a separate color. This is the same chart you viewed when identifying target segments in chapter 9. Here you have the opportunity to customize the chart and save it in a document.

9. Click the Properties button to open the Chart Properties dialog box. Click the Scaling tab, then select the Percent composition button, then click OK.

The width of each bar now represents each segment's value as a percentage of the customer base. Thus, the chart now presents both the Index and Percent (%) Composition of each segment. You may adjust the chart to depict other groups as well. This chart supports the Core and Developmental Group report and illustrates the importance of each segment to Living in the Green Lane.

10. Click the Back button, then select Life Modes as the Target group. Click Next to view the resulting chart. Note that segments are now color coded to represent their Life Mode groups. Repeat this procedure to view Urbanizations groups. The chart should resemble the one below, though the colors may vary. Adjust it to your preferences, then click Next to complete the wizard and create a document displaying the report. Review it, save it if you wish, then close it.

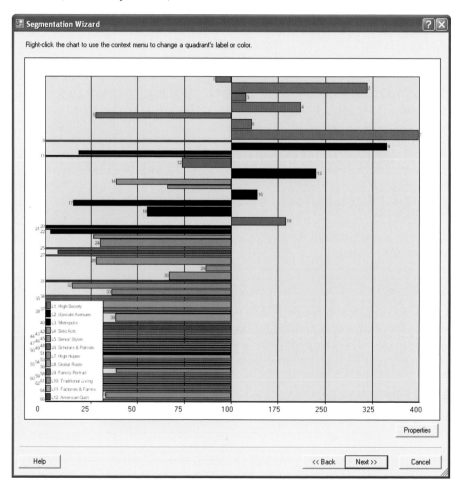

The Segmentation Module provides a Game Plan Chart similar to the one you used to define target segments. You can use it to generate a report describing your Core and Developmental customers.

11. Click the drop-down menu on the Business Analyst toolbar, click Segmentation, then click Segmentation Charts, Maps, Reports. Click Create new segmentation charts, maps and reports, then click Next. Click Segmentation charts, then click Next to open the Segmentation Charts dialog box, which lists the charts available in the module.

12. Select the Four-quadrant (Game Plan) chart, then click Next. In the next window, select LITGL Customer Profile as the Select profile, MinnStPaulCBSA as the base profile and LITGL Target Segments as the Target group. Click Next to open the Game Plan Chart.

    The Game Plan Chart displays the Tapestry Segmentation segments in the same four-quadrant chart you used in chapter 9, with your target segments color coded distinctly. You may adjust the scaling of the chart to define the Niche, Developmental, and Core groups as you wish by changing the axis values.

13. Click the Properties button, then click the Scaling tab. Select each of the options in turn, then click OK to view the impact on the chart. Repeat the process to view each option.

    These options allow you to define the axes as you wish. The options include measures of the number of customers and their purchases with LITGL, which is the volume attribute. The options are:

"Index" & "% Composition"	both of which are based on numbers of customers in each segment
"Index" & "% Composition Volume"	one of which, Index, is based on the number of customers in each segment and one of which, % Composition Value, is based on segment purchases
"Index Volume" & "% Composition Volume"	both of which are based on segment purchases

14. When you have reviewed the options, select "Index" & "% Composition Volume" to display one axis based on number of customers and another based on their purchases. Click OK.

In the fields below the chart, enter **3** in the % Composition Volume field and **100** in the Index field. Your chart should resemble the one below. When it does, click Next.

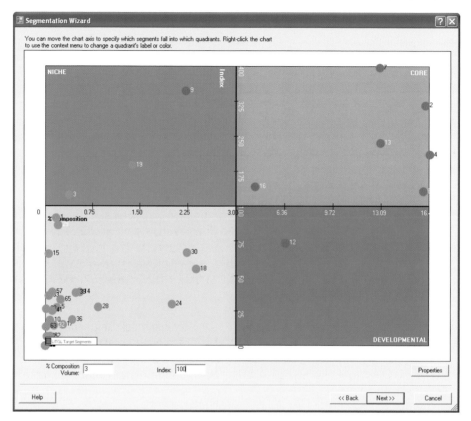

With these settings the group boundaries display both sales and customer information. The two quadrants on the right represent segments that account for three percent or more of customer purchases from Living in the Green Lane. The two segments at the top represent segments that are more prevalent in LITGL's customer base than they are in the general population of the Minneapolis-St. Paul CBSA. Thus, you are identifying customer groups who are both attracted to LITGL and willing to spend money there, a very attractive combination.

The option to revise the axes of the Game Plan Chart allows you to define Core and Developmental groups in the manner most appropriate for your analysis. To remain consistent with the definitions in documentation and report templates, we will revert to our original settings in the following steps. However, remember these options when working with your own enterprise data.

15. Click the Properties button, then the scaling tab. Select the "Index" & "% Composition Volume" option, then click OK. In the fields below the chart, enter **4** in the % Composition field and **110** in the Index field, the default settings, then click Next.

16. In the next window, Enter LITGL Game Plan Chart in the Name and Title fields, select the View Report option, then click Finish to complete the wizard and create a document that displays the chart.

You will use this chart to display the characteristics of your target customers graphically. Now that you have developed a more complete understanding of LITGL's customer base and identified target customer segments, you will create additional reports to help you develop a penetration strategy for your current market.

Create segmentation reports to increase market penetration

The second major objective of segmentation analysis is to increase sales within existing market areas by serving customers more effectively. The Segmentation Module supports this penetration objective with three core reports and maps designed to identify penetration opportunities as well as customer lifestyle, purchasing, and media exposure patterns. The first of these tools is the Market Area and Gap Analysis Report, which identifies opportunities for growth in existing market areas.

1. Click the drop-down menu on the Business Analyst toolbar, click Segmentation, then click Segmentation Charts, Maps, Reports. Click Create new segmentation charts, maps, and reports, then click Next. Click Segmentation reports, then click Next to open the Segmentation Reports dialog box, which lists the reports available in the module.

2. Select the Market Area and Gap Analysis Report, then click Next. In the next window, select LITGL Customer Profile as the Select profile and MinnStPaulCBSA as the base profile, then click Next. In the next window, select the Use thresholds to define groups option, accept the default values of 4 and 110, then click Next.

9

10

3. In the next window, select Block Groups as the geography level and confirm that the Use all features in Block Groups option is not selected. Select LITGL Customers by Store as the customer layer. When the window resembles the one below, click Next.

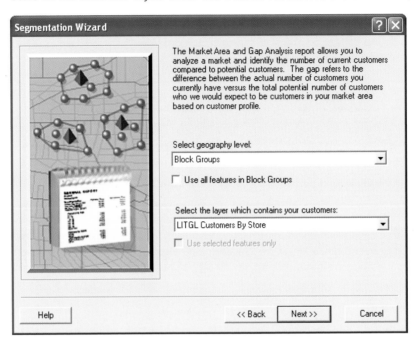

4. In the next window, select the Create Gap Analysis Map option and designate Gap as the thematic map field to be displayed, then click Next. Enter **LITGL Market Area and Gap Analysis** in both the Name and Title fields, select the View Report option, then click Finish.

Using customer layer data relative to Core and Developmental groups, the Segmentation Module calculates the expected number of LITGL customers in each block group where they reside. These values are then displayed as a thematic map and in the Market Area and Gap Analysis report. The report should resemble the one on the next page.

Market Area Gap Analysis

Where are my current Customers and where can I find more?

Now that we know who your customers are and which segments we want to target, we can take a look a where these segments are located in any market.

To analyze a market, we identify the number of current customers, the number of potential Core and Developmental households, and the customer to total household penetration for each standard geograp unit listed below. Using the Tapestry Customer Profile you created, we can also calculate an expected number of customers based on the number of households in a geographic unit and their Tapestry composition.

The gap refers to the difference between the actual number of customers you currently have versus the total potential number of customers who we would expect to be your customer in your market area. To calculate the gap in this market area we have summarized your Customers by the standard geographic units listed below and have estimated how many customers we would expect there to be based on the Tapestry composition of the geographic unit and the Tapestry profile. The Gap is the difference betwee actual customers minus expected customers. The geographies below are sorted by Gap from lowest to highest. Geographies with a large negative gap are under-performing based on the Tapestry profile of area. Geographies that have a large negative gap and contain a large base of potential customers shou be targeted for direct mail applications.

Geographies by penetration and target segment group penetration:

Geo ID	Geo Name	State	Total HHs	%% Core HHs	% Develop HHs	Actual Customer HHs	Customer HH % Pen	Expected Customer	Gap
271230416011	271230416.011	MN	1,707	100	0	1	0.06	6	-5
270190907011	270190907.011	MN	2,626	100	0	4	0.15	8	-4
270530260161	270530260.161	MN	909	100	0	1	0.11	5	-4
271230411051	271230411.051	MN	864	100	0	1	0.12	5	-4
270190906001	270190906.001	MN	2,858	100	0	2	0.07	5	-3
270530260181	270530260.181	MN	1,341	100	0	1	0.07	4	-3
270530260182	270530260.182	MN	806	100	0	1	0.12	4	-3
270530260162	270530260.162	MN	1,057	100	0	1	0.09	4	-3
270530260174	270530260.174	MN	1,023	100	0	2	0.20	5	-3
270530263011	270530263.011	MN	1,526	100	0	4	0.26	7	-3
270530264041	270530264.041	MN	1,601	100	0	4	0.25	7	-3
271410305021	271410305.021	MN	2,291	100	0	1	0.04	4	-3
270530268212	270530268.212	MN	1,401	100	0	1	0.07	4	-3
271230411061	271230411.061	MN	1,039	100	0	3	0.29	6	-3
270530266051	270530266.051	MN	1,264	100	0	2	0.16	4	-2
271230407072	271230407.072	MN	550	100	0	1	0.18	3	-2
270530271011	270530271.011	MN	883	100	0	1	0.11	3	-2
271230408012	271230408.012	MN	921	100	0	3	0.33	5	-2
271230404022	271230404.022	MN	1,064	100	0	2	0.19	4	-2
271630703033	271630703.033	MN	927	100	0	2	0.22	4	-2
270530275041	270530275.041	MN	380	100	0	1	0.26	2	-1
270190905032	270190905.032	MN	629	100	0	1	0.16	2	-1
270530260172	270530260.172	MN	1,156	100	0	3	0.26	4	-1
270530259064	270530259.064	MN	665	100	0	2	0.30	3	-1

The report lists all the block groups in which LITGL customers reside. For each, it lists the number of households in each group, the percentage of these households which are in the Core or Development groups,[1] the number of actual customers, the penetration rate, the number of expected customers, and the gap between the number of expected customers and the number of actual customers. Negative numbers in this column represent opportunities to expand the customer base, while positive numbers represent customer presence above expected levels.

The map now displays the same information, as the layer legend in the Table of Contents indicates. Review the new Gap layer relative to the customer layer, then turn off the customer layer. The map should resemble the one on next page.

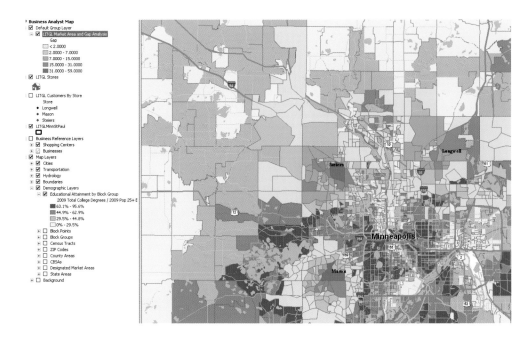

Darker colors represent block groups in which LITGL's penetration is higher than expected, while lighter colors indicate block groups with lower-than-expected penetration. Expected customer counts are based on average penetration rates, which also provide insight into the level of market opportunity available in existing markets.

5. Open the Properties window for the LITGL Market Area and Gap Analysis Report layer. Click the Symbology tab, select Customer Percent (%) Penetration as the Value field, then click OK to symbolize block groups by the percentage penetration of available Core and Development households in each.

    Taken together, these maps reveal that, although LITGL has a substantial customer base in the market area, there is still significant opportunity for increased sales within the area. Two additional reports will help you exploit this opportunity.

6. Click the drop-down menu on the Business Analyst toolbar, click Segmentation, then click Segmentation Charts, Maps, Reports. Click Create new segmentation charts, maps and reports, then click Next. Click Segmentation reports, then click Next to open the Segmentation Reports dialog box, which lists the reports available in the module.

7. Select the Understanding Your Target Customers report, then click Next. In the next window, click the Add button to open the Select MRI Groups dialog box. Press the Ctrl key and click on the Civic Activities, Lawn & Garden, Home Improvement & Services, and Leisure Activities/Lifestyle groups to select them. When your window resembles the one on the next page, click Select to add them to the MRI Groups box. Click Next.

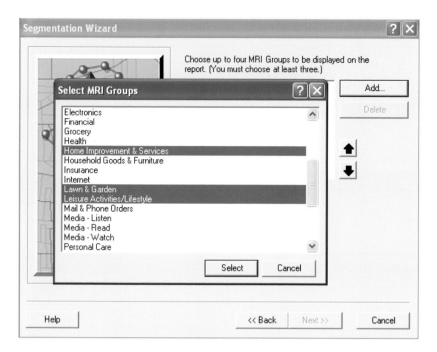

9
10

8. In the next window, select LITGL Customer Profile as the Select profile and MinnStPaulCBSA as the base profile, then click Next. In the next window, select the Use thresholds to define groups option, accept the default values of 4 and 110, then click Next. In the next window, enter **LITGL Understanding Target Customers Report** in both the Name and Title fields, select the View Report option, then click Finish.

The Segmentation Module performs the necessary calculations, generates the report and displays it on the screen, where you may export it to various other formats if you wish. The report includes an explanation of the Indexes as well as listings for each of the designated MRI groups, which resemble this:

Top Civic Activities Characteristics:			
Description	Core Index	Developmental Index	Overall Index
Made a speech in last 12 months	137	103	128
Served on committee for local organization	133	98	126
Written or called a politician in last 12 months	132	102	127
Made contribution to PBS in last 12 months	131	55	127
Engaged in fund raising in last 12 months	130	110	125
Recycled products in last 12 months	126	101	122
Attended public meeting on town or school affairs	125	103	120
Written letter to news/mag editor/called radio/TV	124	95	121
Made contribution to NPR in last 12 months	122	71	121
Voted in federal/state/local election last 12 mo	120	94	116
Participated in environmental grp/cause last 12 mo	119	79	117
Wrote something that has been published last 12 mo	118	66	115
Participated in any public activity in last 12 mo	114	103	112
Member of civic club	108	80	106
Worked for a political party in last 12 months	89	76	95

The values reported here are similar to those you used to profile segments in chapter 6. Here, however, they are aggregated for all the segments in the Core group, the Developmental group, and—in the Overall Index column—for all segments represented in LITGL's customer base.

Recall that index values relate to the base profile average for a particular item, so that an index value of 140 for the Core Index, for example, means that respondents in this group are 40 percent more likely to report this behavior than the Minneapolis-St. Paul average. To facilitate interpretation, the indexes are sorted in the descending order in the Core Index column. Further, if an MRI group contains more than 20 items, only the top 20 are listed in the report.

This report is helpful in two ways. First, it reflects purchasing patterns for a company's goods and services. In this case, the Lawn & Garden values, as well as the Home Improvement & Services indexes, indicate the goods and services within LITGL's product and service mix most often purchased by the various customer groups. This information is very useful in determining product and merchandising strategy, as you learned in chapter 6.

Second, the report also reflects lifestyle characteristics of a company's target customers. In this case, the Civic Activities and Leisure Activities/Lifestyle indexes report the frequency with which LITGL's customer groups engage in these activities relative to the average of the Minneapolis-St. Paul CBSA. This helps the company understand the lifestyle patterns of its customers and the role of the company and its products within that lifestyle. These insights are useful in merchandising and store layout decisions as well as in crafting promotional messages attractive to target customers.

If you wish, rerun this report and select different MRI groups to learn more about Living in the Green Lane's customers.

With these additional insights into its customer base, you are ready to refine your penetration strategies for this market area. To do so successfully, you must communicate effectively with target customers. The Developing Marketing Strategies Report will help you identify the most appropriate media for doing so.

9. Click the drop-down menu on the Business Analyst toolbar, click Segmentation, then click Segmentation Charts, Maps, Reports. Click Create new segmentation charts, maps and reports, then click Next. Click Segmentation reports, then click Next to open the Segmentation Reports dialog box, which lists the reports available in the module.

10. Select the Developing Marketing Strategies Report, then click Next. In the next window, select LITGL Customer Profile as the Select Target profile and MinnStPaulCBSA as the Select Base profile, then click Next. In the next window, select the Use thresholds to define groups option, accept the default values of 4 and 110, then click Next. In the next window, enter **LITGL Developing Marketing Strategies Report** in both the Name and Title fields, select the View Report option, then click Finish.

The Segmentation Module performs the necessary calculations, generates the Developing Marketing Strategies Report and displays it on the screen, allowing you to export it to the data format of your choice.

The layout and organization of this report is identical to the Understanding Target Customers Report displayed above. The difference is that it identifies the top 20 media vehicles that customers in the Core group read, watch, or listen to. It also reports the indexes for Developmental customers and the overall customer base for these media vehicles.

This report is very useful in identifying the best media vehicles for communicating with Living in the Green Lane's customer base. In this role, it makes a vital contribution to the company's promotional campaigns.

Taken together, these reports allow you to assess the potential for increased penetration of the current market area, understand the products customers purchase, and learn how they fit into their lifestyle patterns and how to communicate LITGL's promotional messages to them more effectively.

9

10

Create segmentation reports to identify expansion opportunities

Note: If you are using the Minnesota dataset from the Media Kit accompanying this book, you will *not* be able to perform the following steps as you lack the data license to do so. However, you should follow the process and observe the steps to understand how the Segmentation Module uses the data from a local segmentation study to identify expansion opportunities at the national level. If you are using a Business Analyst Desktop and the Segmentation Module with a full national data license, you will be able to perform each of the steps in this process as they are described.

The third key objective of segmentation analysis is to provide Living in the Green Lane a much more substantial growth opportunity to expand to other markets in the United States with its green-lifestyle center concept and extensive line of green-lifestyle products and services. The Segmentation Module supports that objective with reports and maps designed to identify concentrations of population with characteristics matching those of LITGL's best customers. You will use those tools to assess the potential for geographic expansion.

As you begin, recall Janice and Steven's expansion strategy. Based on their experience in the Minneapolis-St. Paul area, they wish to identify other CBSAs in the United States that can support four LITGL locations. Ideally, at least two of the stores would be company owned, though the remainder could be franchised to appropriate partners. To build a company store, Janice and Steven require a 3-mile ring service area with home-improvement material sales of more than $15 million per year. Market areas below that level but above $10 million would be candidates for franchise agreements. You will use the tools of the Segmentation Module to identify these opportunities.

1. Click File, click Open. Navigate to C:\My Output Data\Projects\LITGL Minneapolis St Paul\CustomData\ChapterFiles\Chapter10\LITGLSegmentation2.mxd. Click the map file to open it. When prompted to save your existing file, do so, saving it with a different file name in the same folder.

    This map is the standard Business Analyst map, with the addition of a thematic layer depicting Home Related Expenditures by Household at the block group level in the Minneapolis-St. Paul CBSA. At this scale, this layer is discernible only as a gray polygon covering the Minneapolis-St. Paul area. You will use this national map to identify attractive areas for LITGL's expansion strategy.

    The Segmentation Module offers three reports to support market expansion analysis. You will begin by using the Market Potential Volume Report to identify CBSAs in the United States with high sales potential for Living in the Green Lane. Another relevant report, The Market Potential Report, would be used for this analysis in the absence of volume data, which it lacks.

2. Click the drop-down menu on the Business Analyst toolbar, click Segmentation, then click Segmentation Charts, Maps, Reports. Click Create new segmentation charts, maps and reports, then click Next. Click Segmentation reports, then click Next to open the Segmentation Reports dialog box, which lists the reports available in the module.

3. Select the Market Potential Volume Report, then click Next. In the next window, select LITGL Customer Profile as the Select profile, MinnStPaulCBSA as the base profile and Total Households as the Base, then click Next. In the next window, select CBSAs as the Geography level, then click Next.

4. In the next window, select the Create Market Potential Map option and designate Expected Volume as the Thematic Map field, then click Next. In the next window, enter **LITGL CBSA Market Potential Report** in both the Name and Title fields, select the View Report option, then click Finish.

    Based on LITGL's customer profile, the Segmentation Module calculates various market size measures for CBSAs in the United States and creates a report which lists them. In addition, it creates a map layer that symbolizes the expected volume of the CBSAs. The report should resemble the one below,[2] and include many of the same attributes as those you have developed in your market penetration analyses. Review it to identify CBSAs with the largest expected volume and average expected sales per household.

Market Potential Volume Report

Where can I find potential Customers?

Market Potential data measures the likely demand for a product or service for your market area by a specific geography level. You can use this report to make decisions about where to offer products and services.

The Expected is the estimated number of adults or households that use a particular product or service.

The Penetration Percent is a measure of the percent of adults or households that use a particular product or service compared to the Total Households or Total Adults in the geography.

The Index measures the likelihood of adults or households in a specified area to exhibit certain consumer behavior compared to the base area average. The index is tabulated to represent a value of 100 as the overall demand for the base area. A value of more than 100 represents high demand; a value of less than 100 represents low demand. For example, an index of 120 implies that demand in the trade area is likely to be 20 percent higher than the base area average; an index of 85 implies demand is 15 percent lower than the base area average.

The Expected Volume is the estimated volume usage of a particular product or service. Depending on the volumetric value used, this could be a count or a dollar amount.

The Average Volume is the average volume usage per adult or household.

The Volume Index measures the likelihood of adults or households in a specified area to exhibit certain consumer behaviour compared to the base profile average. The index is tabulated to represent a value of 100 as the overall demand for the base area. A value of more than 100 represents high demand; a value of less than 100 represents low demand. For example, an index of 120 implies that demand in the trade area is likely to be 20 percent higher than the base profile average; an index of 85 implies demand is 15 percent lower than the base profile average.

Geographies by Market Potential Data:

Geography ID	Name	Total Households	Expected Households	Percent Penetration	Index	Expected Volume	Average Volume	Volume Index
20940	El Centro, CA Metropolitan St	49,511	19	0.0	24	203,448.03	10,707.79	105
25620	Hattiesburg, MS Metropolitan	53,215	22	0.0	26	233,361.64	10,607.35	104
33220	Midland, MI Micropolitan Stati	33,212	28	0.1	54	293,868.81	10,495.31	103
26700	Huron, SD Micropolitan Statis	6,967	1	0.0	9	10,561.83	10,561.83	103
32620	McComb, MS Micropolitan St	20,973	1	0.0	3	10,561.83	10,561.83	103
31700	Manchester-Nashua, NH Met	153,735	234	0.2	97	2,437,591.22	10,417.06	102
49340	Worcester, MA Metropolitan	299,726	353	0.1	75	3,685,415.70	10,440.27	102
45000	Susanville, CA Micropolitan S	10,151	9	0.1	56	93,384.05	10,376.01	102
30580	Liberal, KS Micropolitan Stati	7,559	3	0.0	25	31,128.02	10,376.01	102
12460	Bainbridge, GA Micropolitan	11,050	4	0.0	23	41,504.02	10,376.01	102
22340	Fitzgerald, GA Micropolitan S	10,527	3	0.0	18	31,128.02	10,376.01	102

The map should resemble the one on the next page.

CBSAs in darker colors are those with the highest levels of expected sales volume, while those in lighter colors have lower values. Several CBSAs are candidates for LITGL's initial geographic expansion.

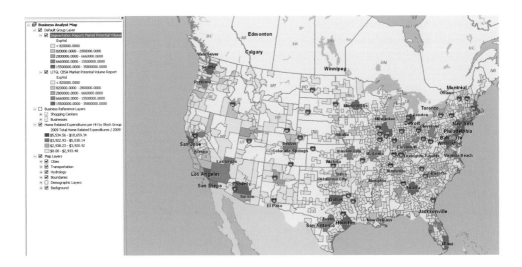

Based on the report and map, Janice and Steven have chosen to explore the Seattle-Tacoma-Bellevue, Washington, CBSA first, as it has approximately the same number of households as LITGL's current market with similar market conditions and sales opportunities. In addition, this region of the country is known for its concern for environmental and sustainability issues, rendering it an attractive candidate for expansion.

5.   Zoom to the Seattle-Tacoma-Bellevue CBSA. Note that the Olympia, Washington, CBSA is contiguous to Seattle and will also be included in the analysis. Run a second Market Potential Volume report, but select Census Tracts instead of CBSAs as the Geography level and enter **LITGL Seattle Census Tract Market Potential Volume Report** as the Name and File fields. Select View Report in the final window, then click Finish.

The Segmentation Module repeats its processing steps and produces a similar report and map. The report layout is similar to the one above, though it reports data at the census tract, rather than the CBSA level. It may require editing with Crystal Reports to display all fields properly. The map should resemble the one on the next page.

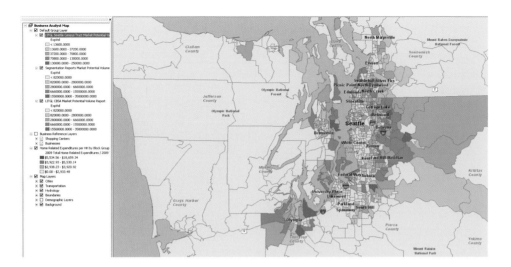

This map offers a more detailed picture of expected sales concentrations in this market area. Darker colors represent higher levels of expected sales, lighter colors represent lower levels. You will use this map to assess the ability of this market area to support multiple LITGL stores. Before doing so, however, you will explore the distribution of the company's Core, Developmental, and Niche groups in this area with the Four-Quadrant (Game Plan) Map.

The Game Plan map is one of Segmentation Module's maps that support the analysis of growth opportunities. The others are the Target Map and the Target Penetration Map. The Target Map reports the presence or absence of target customer segments in designated levels of geography. The Target Penetration Map reports the penetration levels of target customers as a percentage of total households by selected geography. These are useful measures of concentration when sales data is lacking, but that is not the case here.

6. Click the drop-down menu on the Business Analyst toolbar, click Segmentation, then click Segmentation Charts, Maps, Reports. Click Create new segmentation charts, maps, and reports, then click Next. Click Segmentation maps, then click Next. In the next window, select the Four Quadrant (Game Plan) Map option, then click Next.

7. In the next window, select the Create new Game Plan chart option, then click Next. In the next window, select LITGL Target Segments in the Groups field, LITGL Customer Profile as the Target profile and MinnStPaulCBSA as the base profile. The next window displays the Game Plan chart. Accept the default values of 4 in the Percent (%) Composition field and 110 in the Index field, then click Next.

8. In the next window, select Tracts as the Mapping Layer and Total Households as the Segmentation base. Select the mapping layer option but NOT the mapping report option, then click Next. Enter **LITGL Seattle CT Game Plan Map** in the Name field, then click Finish. When the map layer is displayed, open its Properties box, click Display, and set the Transparency of the layer to 0% so its values are clear.

The Segmentation Module assesses the composition of each census tract in the map extent, assigns it to one of four quadrants in the Game Plan model, and creates a map layer which indicates the designation of each census tract. The map should resemble the one below.

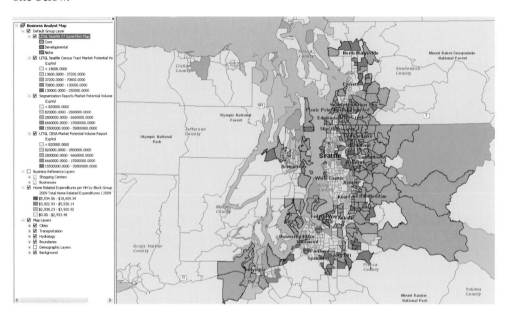

This map identifies concentrations of LITGL's Core, Developmental, and Niche customers. Core customer concentrations present the strongest expansion potential, though Developmental and Niche customers are also important for future growth. Toggle this map with the Market Potential Volume layer to compare customer segments with anticipated sales levels.

These maps display the distribution of expected sales and customer segments across this potential market area. You must now determine whether or not this distribution will support four new stores, which is Janice and Steven's objective. You will use the Dynamic Ring tool, which you also employed in chapter 3, for this purpose.

9. Turn off the Game Plan layer. Click the Dynamic Ring Analysis Tool 🔭. This loads a graph displaying selected attributes on the left of the map. Click the Change Parameters button at the bottom of the graph to load the Dynamic Ring Analysis wizard. Adjust the settings of the wizard until it resembles the screen on the next page, then click Next.

10. Click Next. In the resulting box, select the threshold option, select Home Imp Material-Own & Rent: Tot as the threshold field, and enter **15000000** ($15 million) as the threshold value. Set the Radius as 3 and the Distance unit as Miles. With these settings the Dynamic Ring tool will distinguish between those 3-mile-ring areas in which total home improvement materials expenditures are greater than $15 million and those that fall below this level. Recall that this is the level of sales Janice and Steven require for a market area to support a company-owned store. This box should resemble the one below. When it does, click Finish to apply the settings and close the box.

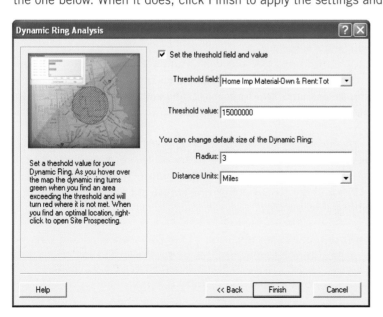

11. With the Dynamic Ring icon, move around the map. Locate a site near dark green census tracts with high levels of expected sales volume and click on it. Depending on where you clicked, your map will resemble this one in structure, but not in location.

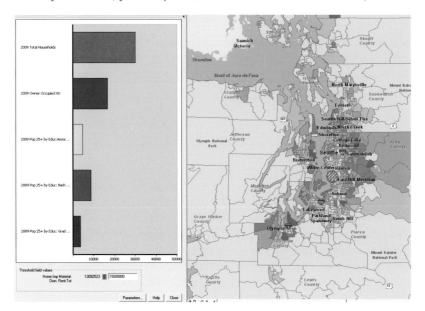

The map displays a hashed circle in a 3-mile ring around your chosen site. A red ring indicates that this area falls below the $15 million expenditure threshold, while a green ring indicates an area above the threshold. The bar chart to the left displays the number of households, owner-occupied housing units, and persons with associate, bachelor, and graduate degrees in that area. Just below the chart is a field which reports the expected sales value within this ring and whether this value exceeds (green square) or is exceeded by (red square) the threshold sales level.

Continue to move around the potential market area, trying several locations in search of attractive potential sites. If you find four geographically dispersed sites that meet the expenditure threshold, Living in the Green Lane can serve the market area with its own stores.

If you do not find four such sites, or if you wish to explore the incremental expansion potential of franchising, click the Change Parameters button and reduce the threshold from $15 million to $10 million. Recall that this is Janice and Steven's expenditure threshold for franchising agreements.

12. Revisit several of the sites that did not meet the $15 million threshold and click again. Note that the actual expenditure figure is reported in the box at the bottom of the bar chart window. Use this procedure to identify potential sites whose expenditures lie between $10 million and $15 million. These are candidates for expansion through franchise agreements.

If you identify at least two sites that meet the threshold for company-owned stores and at least two other sites that meet either this threshold or the lower threshold for franchise agreements, you may conclude that the Seattle-Tacoma-Bellevue, Washington, area will support Living in the Green Lane's objectives for its initial geographic expansion. You may repeat this process with other attractive CBSA's to develop a national expansion plan. You will then be ready to replicate the trade area and customer analytical procedures you used to select specific store sites in the Minneapolis-St. Paul area in each of the new CBSA's. Does it surprise you that Business Analyst Desktop provides the analytical tools and data you will need for that analysis as well?

Create segmentation study report

For purposes of formally reporting segmentation analysis results, the Segmentation Module provides the capability of generating a Segmentation Study report that integrates the results of several of the analyses you have performed. The wizard that creates the report uses a template that combines descriptive text with report elements based on several segmentation tools. You may revise the default settings to include segmentation reports of your choice as well as reports generated by other Business Analyst Desktop procedures. To use the wizard, you identify the existing reports you wish to include and provide appropriate settings for each new report element.

1. Click the drop-down menu on the Business Analyst toolbar, click Segmentation, then click Segmentation Study. Click Create a new Study, then click Next. In the next window, click Standard Segmentation Study Template, then Next to open the Segmentation Study Contents page, which looks like this. (If a box appears informing you that Some items are not valid, click OK to continue. Repeat as necessary throughout this process.)

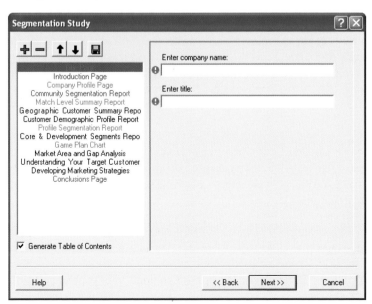

2. Enter **Living in the Green Lane** in the company name box and **Living in the Green Lane Segmentation Study** in the title box. Click the Company Profile Page element in the left box, to open its dialog box. Click the Remove button ▬ at the top of the box to remove it from the list, as you do not wish to include a company profile.

3. Click the Match Level Summary Report element to open its box. Select LITGL Customers by Store as the customer layer and enter **LITGL customers and their store affiliation** in the description field.

    Note that as you complete the required fields for an element, its color changes from red to black in the elements list. You should review the elements in black type as well to observe the settings with which they will be processed. When all the elements are complete you are ready to run the study.

4. Continue to supply required information for each of the elements in the list. The following appear in many of the elements and should be entered consistently:

    Geography level—Block Groups
    Target segmentation profile—LITGL Customer Profile
    Base segmentation profile—MinnStPaulCBSA
    Percent composition threshold—4
    Index threshold—110

    When prompted for descriptions of fields, enter brief descriptions of your choice.

5. When the required fields have been entered for all the elements and they are listed in black, click Next. In the next box, enter **LITGL Segmentation Study** as the Study name, confirm that the Print this Study now! option is *not* selected, then click Finish.

Using the settings you provided, the Segmentation Module generates all the study elements, combines them into a single document, and displays the Study on the screen, allowing you to export it to a variety of data formats. The study is quite long, but the title page should resemble this:

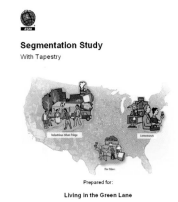

Segmentation Study
With Tapestry

Prepared for:
Living in the Green Lane

The formal Segmentation Study is now ready to be printed to serve as a support document for your presentation, conclusions, and recommendations. By adjusting the contents of the study to the exact topic of your analysis, you can customize it to the exact reporting requirements of each project.

In this chapter you used the profiles you created in chapter 9 to produce segmentation reports, maps, and charts that allowed you to target Living in the Green Lane's customer base more precisely, penetrate current markets more effectively, and identify growth opportunities more accurately. These activities mark the evolution of LITGL from a small, born spatial enterprise into a mature organization with realistic plans for national expansion.

Business Analyst Desktop has been the core analytic technology in that process and will continue to be crucial to the company's marketing and growth strategies. For this reason, it is imperative that you integrate Business Analyst's capabilities throughout the organization. In the next chapter you will turn to that task with Business Analyst Server.

ROI considerations

The analyses supported by the Segmentation Module provide two streams of incremental revenue, market penetration (which generates revenue from increased sales in existing market areas), and market expansion (which generates revenue from sales in new market areas).

Market penetration produces increased revenue with higher sales levels in existing market areas. These incremental sales result from better understanding of the existing customer base, which results in more effective merchandising, marketing, and promotional programs to these customers and similar prospects in existing market areas.

By identifying Core, Developmental, and Niche target groups, companies determine more precisely the identity of their best customers. Market area gap reports identify concentrations of the segments that comprise these groups and indicate which geographic areas present opportunities for greater penetration. Reports that contain significant Market Potential Index values for these target groups help companies shape their marketing and promotional programs to exploit these opportunities more effectively.

Integration of customer purchasing data with profiling and lifestyle information within the Market Area and Gap Analysis provides the foundation for estimating the incremental revenue potential of effective market penetration.

In this case, the Market Area and Gap Analysis report you produced in chapter 10 indicated that there are 77 fewer customers than expected in block groups assigned to the core target group. As these are the segments most attracted to Living in the Green Lane's business model, it is reasonable to conclude that more effective marketing and promotional estimates could convert half of these prospects into new Living Green members, resulting in 38 new customers. Further, if the annual purchases of the new members are half the $10,172 average for all Green Living Club members, the resulting incremental revenue is approximately $193,268. Note that these increases are not based on increases in marketing or promotion costs but rather better allocation of existing resources to these activities.

This conservative estimate of incremental revenue does not include increased sales to other existing customers, potential new customers from the Developmental and Niche target groups, or new customers from block groups with no penetration gap.

The second stream of incremental revenue—market expansion—produces sales from new company-owned stores and franchise operations in new geographic regions. As these operations would produce some new revenue wherever they are located, the relevant figure for this estimate is not total new sales dollars, but the incremental dollars resulting from a good expansion choice versus a poor one. The CBSA Market Potential Volume Report provides this data.

Janice and Steven based their choice of the Seattle area on its similar size and sales potential relative to the Minneapolis-St. Paul area. Absent the volume estimates of the Segmentation Module, they would have been limited to demographic comparisons such as numbers of households, income levels, etc. Of the six CBSAs closest to Minneapolis-St. Paul in number of households, the Seattle area has the highest expected sales volume for Living in the Green Lane at $18,868,400, with the others ranging from $8,984,208 to $17,628,800. Thus, the incremental revenue of selecting the Seattle region ranges from $1,239,600 to $9,884,192 with an average of $6,148,307.

You may recall that these figures take into account only sales to Green Living Club members, who are the base group for LITGL's customer profile. Additional sales to nonmembers would significantly increase that figure.

In sum, the market and penetration and market expansion initiatives supported by the Segmentation Module produce significant projected incremental revenue. By contrast, the incremental costs include only the analyst's time in developing the charts, maps, and reports used to make these decisions as you have used the Segmentation Module to create block group level analyses and profiles in previous chapters.

When these strategies are implemented, Living in the Green Lane will evolve from a successful regional company to a growing national chain of green-lifestyle centers. As it does, it will be vital to integrate Business Analyst's data and tools throughout the information technology infrastructure of the enterprise. In the next chapter, you will use the Business Analyst Server to accomplish this objective. Before doing so, however, pause for a moment to reflect on what you have accomplished in Part VI.

You have expanded your GIS knowledge by learning:

1. The value of geocoding customer records and attaching data attributes to them based on their location
2. The value of developing customer profiles from customer lists and sales data
3. The value of comparing customer and population profiles to identify high value customers
4. The benefits of using customer lifestyle, purchasing, and media exposure information to craft marketing strategies for serving them better
5. The usefulness of high-value customer profiles in assessing geographic expansion opportunities
6. The value of GIS maps, reports, charts, and segmentation studies in communicating the results, conclusions, and recommendations of research projects

You have enhanced your GIS skills by using the Business Analyst Desktop Segmentation Module extension to:

1. Geocode customer addresses and attach demographic and segment data to them
2. Develop profiles from customer lists, layers, geographic areas, and survey data
3. Develop reports which summarize customer profile characteristics
4. Assign segments to Core, Developmental, and Niche target groups based on their composition and/or sales volume
5. Create reports using Market Potential Indexes to determine target group behaviors, values, lifestyles, purchasing patterns, and media exposure
6. Assess the potential sales volume of geographic areas based on their similarity with attractive customer profiles
7. Assess the ability of attractive geographic areas to support new company-owned and/or franchised stores
8. Create customized formal Segmentation Studies to support research conclusions and recommendations

Notes

1. Note that the all the values for percentage of Core and Development groups are either 0 or 100. As designations are made at the block group level, by definition all the households in a block group are assigned to the same segment. In larger geographic units, these percentages would be based on several diverse designations and would display more variation.

2. The report in the image has been edited with Crystal Reports to display all fields and appear in descending order of Volume Index, then Index. In your report, the report fields with ###### indicate values that are too large for the report field. If you wish, you can use Crystal Reports to edit the report design, correct this formatting weakness, and sort features in the order you wish.

Sharing Business Analyst resources across the enterprise with Business Analyst Server

Relevance	Businesses wishing to integrate business GIS functionality into their managerial and decision-making processes by making Business Analyst tools available to managers through a browser-based interface.
Business scenario	LITGL's expansion plans have succeeded and the company has identified additional opportunities to grow through geographic expansion, additional company-owned centers, and franchise operations. The company must now redesign its organizational structure to support this accelerated growth pattern while maintaining managerial effectiveness and responsiveness to local market conditions.
Analysis required	LITGL must develop an enterprise approach to integrated business GIS management that will support its new organizational structure, make business GIS functionality available to managers at all levels, and match business GIS maps, analyses, reports, and tools to the responsibilities and capabilities of managers at various levels of the organization.
Role of business GIS in analysis	To achieve these objectives, LITGL must implement a system that allows system administrators to: • Design customized maps and expose them to users as map services. • Create Business Analyst projects and expose them to users in Business Analyst Server applications. • Create a system of users, roles, and enabled tasks that matches each management level with the appropriate business GIS resources. • Manage existing workflows and design new ones to guide users systematically through business GIS research projects, and allow managers using the system to: • Access the business GIS tools appropriate for their roles and responsibilities in the enterprise. • Integrate their own enterprise data into business GIS analyses. • Use business GIS tasks and workflows required for their analyses. • Use the maps, reports, charts, and information generated by their analyses in decision making and managerial processes.
Integrated business GIS tool	ESRI Business Analyst Server.
ROI considerations: Cost of business GIS	Software and server-acquisition costs, maintenance costs for server, training costs for system administrator.

ROI considerations: benefits of business GIS	Dissemination of integrated business GIS functionality to managers throughout the enterprise with acquisition costs for software, hardware, and personnel considerably lower than multiple installations of single systems.
Environmental impact of business decision	The environmental impacts identified in previous projects would be realized throughout the enterprise with the dissemination of integrated business GIS tools.

Table VII.1 Executive summary

The Living in the Green Lane scenario

Living in the Green Lane's expansion plans have been successful. The company has entered six CBSAs across the United States. In those markets, LITGL has opened 10 company-owned stores and has franchised an additional six units. The site selection methods and customer analytics you have performed to this point have been extended to these operations effectively, making a significant contribution to their success. These analyses indicate that several additional markets offer the company favorable expansion opportunities, a conclusion supported by the large and increasing number of inquiries from potential franchises in new markets.

Janice and Steven wish to exploit these opportunities and expand the pace of LITGL's geographic expansion. However, they recognize that the current organizational structure limits their capacity to do so. To date, they have been actively involved in each expansion decision, relying to a large extent on the integrated business GIS tools with which you have evaluated each expansion decision.

To increase the pace of expansion and manage the rapidly growing enterprise, Janice and Steven wish to reorganize the enterprise to a more decentralized, nimble, and flexible structure. Specifically, they wish to create a system of four regional managers, each responsible for overseeing existing centers and opening new ones in his/her region of responsibility. Living in the Green Lane's home office will continue to have primary responsibility for prioritizing expansion opportunities nationwide at the CBSA level. However, the regional managers will be charged with overseeing existing stores and opening new ones within the designated CBSAs in their regions.

Janice and Steven also wish to concentrate significant responsibility at the store manager level. Consistent with their concept of service to local market areas, these managers will be empowered to adjust their product and service mixes, merchandising tactics, and marketing strategies to the market areas they serve. Although they are responsible to regional managers for sales, profit, and customer satisfaction, they will have considerable autonomy to select the best strategies for meeting required performance benchmarks.

Based on their experience, Janice and Steven recognize that this reorganization requires a new approach to integrated business GIS. While centralized analysis has served the company well to date, it is not sufficient to serve the rapidly growing needs of the new organization. Moreover, decentralized decision making in the new structure will require a similar pattern in the management of integrated business GIS systems. This is especially true for LITGL because Janice and Steven wish to develop business GIS skills as core competencies of the company's management team. They want managers to work directly with business GIS tools rather than reports and maps ordered from a central office. They believe that this approach will result in more intuitive, creative uses of the technology by executives throughout the enterprise and result in even greater leverage of LITGL's investment in integrated business GIS.

This philosophy and the new organizational structure require you to manage business GIS systems in a way that makes them accessible to all levels of the management team, encourages their use, and provides learning opportunities for new managers encountering them for the first time. At the same time, you must ensure the integrity of systems data, protect enterprise data sources, maintain the validity of enterprise analytical procedures, and present business GIS capabilities to executives in a manageable, relevant user environment.

This is a complex set of objectives, but it is essential to establishing business GIS as a core decision support tool throughout the enterprise. In chapter 11, you will learn how the Business Analyst Server system can help to achieve them.

Extending integrated business GIS tools with Business Analyst Server

The ESRI Business Analyst Server system is composed of several layers of business GIS functionality designed to move Business Analyst tools from the desktop to an enterprise-wide platform. It allows analysts to develop customized procedures and deliver them to the relevant managers in the enterprise. This, in turn, allows managers with relatively modest business GIS skills to exploit the potential of the ESRI Business Analyst suite.

The first level of this system is Business Analyst Desktop and its Segmentation Module extension. Systems administrators can use these resources to create customized datasets, maps, and analyses for use by managers across LITGL's operations. These are the same tools and procedures you have used to manage the company's growth to this point, but they focus on the geographic areas to which the company has expanded or is contemplating expansion.

You will expose these resources to managers in the organization in two formats. The first is map files from which you will create map services to be included in Business Analyst Server projects. While managers will consume these services in the Business Analyst Server context, you may also use them to create Web applications that would be available to LITGL's customers through their Web browser and/or ArcGIS Explorer, a free map-reading utility provided by ESRI. The system also offers integration possibilities with Google and Microsoft map services.

The second format is Business Analyst Server project files, which can integrate the map services you create with analyses you have performed using Business Analyst and include in project files. You may upload these project files to Business Analyst Server to make them available to managers who require the information contained in your analyses.

The next component of this system is ArcGIS Server, which facilitates the process of serving maps over the Web, designing applications to present them to users, and implementing security systems that manage user access to maps, data, and spatial procedures.

Specifically, you will use ArcGIS Server to create map services from the map files you design for LITGL's managers. You will also use the ArcGIS Server security system to create user names for potential users, generate roles that define user privileges within the system, and match user groups and roles to specify the access levels of each user group. In LITGL's case, users in the regional manager role will have access to the site selection functions in Business Analyst Server, while those in the store manager role will have access to customer profiling functions. You will also create a new store manager role focused on the tasks with which new store managers will develop Tapestry Segmentation Profiles of their customers.

Business Analyst Server is the final component in the system. Its function is to integrate the functions of Business Analyst Desktop and ArcGIS Server in order to disseminate Business Analyst functionality throughout the enterprise. It does so by integrating the functions of these systems and expanding their capability. For example, Business Analyst Server not only integrates map services developed on the desktop into projects, but also adds to them the extensive collection of layers contained in the system's default maps.

Similarly, Business Analyst Server incorporates the analyses and reports, along with a map, built into uploaded project files while also allowing users to execute Business Analyst procedures of their own within the context of their browser interface.

In addition, Business Analyst Server enables system administrators to manage workflows to use with the system. Workflows are sequences of tasks that systematically produce desired analytical outcomes. Business Analyst Server includes two workflows: Customer Profiling and Site Selection Evaluation. Each walks users through the processes of loading enterprise data, integrating it with Business Analyst data, designing trade areas and/or customer pro-files, and producing the reports and maps that communicate the results to managers. System administrators can also create new workflows customized to the needs of the enterprise and/ or distinct user groups.

Finally, Business Analyst Server incorporates the security system of ArcGIS Server, but extends it by allowing system administrators to determine the collection of Business Analyst tools available in a Web environment for each user role. This security system provides the capability to match users with the Business Analyst tools most relevant to their information requirements. It also provides the capability of matching users' access to business GIS tools with their knowledge and skills in the field. These capabilities also allow administrators to use the system as a training tool, making basic functions available to new users and expanding the range of available tools as users gain skills and are assigned to new roles in the system.

Thus, by integrating these components, Business Analyst Server provides a comprehensive system for disseminating business GIS capabilities throughout the enterprise, integrating enterprise and Business Analyst data, protecting enterprise data resources, and matching user capabilities and responsibilities with appropriate integrated business GIS tools. In chapter 11, you will explore this system from the perspectives of a systems administrator designing applications, and a new store manager creating a Tapestry Segmentation Profile for the LITGL center in St. Louis.

Chapter 11

Serving Business Analyst applications with ESRI Business Analyst Server

ESRI Business Analyst Server is a complex, comprehensive analytical system that integrates its functions with ESRI Business Analyst Desktop, including the Segmentation Module extension, and ArcGIS Server to produce a customized, online, integrated business GIS system. It requires an extensive installation process and access to enterprise information technology resources that cannot be easily replicated by the software accompanying this book.

For this reason, chapter 11 takes a descriptive rather than interactive "hands on" approach to presenting Business Analyst Server functionality. Specifically, it presents the process first from the role of a system administrator, the role you would fill as LITGL's director of business GIS applications, and then from the role of a new store manager using the system for the first time.

The system administrator perspective will be presented via a description of how to create a basic Business Analyst Server application that provides new managers the ability to explore maps of their competitive environment and to produce a Tapestry Segmentation Profile of their customers. You will observe, but not actually perform, the steps with which this is accomplished.

From the user perspective, you likewise will read a description of how to sign on in your role as a new store manager, view Business Analyst maps of your market area, upload customer data to the system, create a Tapestry Segmentation Profile of your customers, and generate maps and reports for use in deciding how best to serve this customer base.

This exercise will provide an overview of Business Analyst Server's potential, but will cover only a fraction of its capabilities. It will nonetheless illustrate effectively the process of moving from Business Analyst tools on the desktop to an enterprise-wide system for enabling integrated business GIS analyses by managers based on their roles and responsibilities.

Managing BA Server functionality as system administrator

You will begin by following the procedures through which a system administrator designs and serves a Business Analyst application to users throughout the enterprise.

Design a Business Analyst map, use it to create an ArcGIS map service, and add it to Business Analyst Server

The process of creating a Business Analyst Server application begins with designing a Business Analyst map for the relevant market area. In this instance, that market is the St. Louis CBSA. As system administrator you will design a basemap for this CBSA similar to those you have been working with in the Minneapolis-St. Paul area, but displaying data for the St. Louis market area. You will use the Business Analyst Desktop installation on your server to create this basemap.

The map below illustrates this process. It includes some of the same data layers you used in earlier chapters, but displays them for the St. Louis CBSA. Living in the Green Lane's St. Louis store is displayed, as are competing home centers and area shopping centers. In addition, the Table of Contents contains thematic layers for the key characteristics of LITGL's best customers. These data layers are based on standardized Business Analyst data. To publish this file as a map service, you first will save it to the active project folder as LITGLStLouis.mxd.

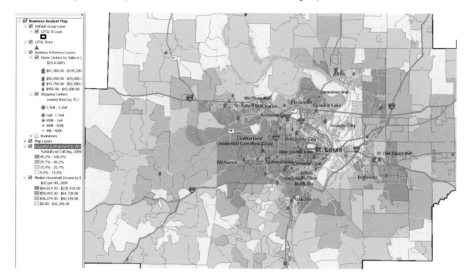

To allow access to this map in Business Analyst Server, you must make it available as a map service using ArcGIS Server. A map service is a map document that is available to users through a Web interface. Users may use browser technology, GIS clients such as ArcGIS Explorer, and/or desktop GIS applications to access these services. Thus, map services provide an excellent means of providing access to dispersed users by exploiting Internet delivery methods. To do so, you will run ArcGIS Server Manager on the host server and log in with an administrator account. After doing so, you will reach the screen below.[1]

11

This screen allows you to perform several functions in ArcGIS Server. You will use it again at several points in this project. For this task, you select the option to Publish a map, globe, or other GIS resource as a service. This function takes a GIS resource, such as the Business Analyst map you just created, and make it available to other users in a Web interface such as a map service, i.e., as a map document available through Internet technologies.

Before proceeding with this task, note the Create a web application option. Selecting this option opens a wizard that allows you to select a map service such as LITGLStLouis.mxd and use it to create a full-blown Web site that combines the map with a collection of GIS tools, links, functions, and display options of your choice. As you will see below, you can add significant additional functionality to the site by designing it in Business Analyst Server. When you select the option to create a service from a GIS resource, you will reach the window below, in which you identify the resource to publish, in this case the LITGLStLouis. mxd map file, and assign it a name, also LITGLStLouis. The screen resembles this:

By clicking Next through two additional screens to accept the default values, you will complete this task and publish LITGLStLouis.mxd as a map service.

When this process is completed, the ArcGIS Server returns you to the Manage Services window, where all active services are listed. The new service appears on the list, indicating that it is ready for use in ArcGIS Server and Business Analyst Server applications. As a map service, it is also available for authorized users to access directly from ArcCatalog, ArcMap, and ArcGIS Explorer or for developers to include in enterprise Web sites with the ArcGIS API for Javascript.[2]

To add the map service to Business Analyst Server, you will run the Business Analyst Web Applications Post Install routine on the server. You will run the Set Analysis Center Map Services option and then select Default Map as the required map service with Business Analyst capabilities. The next window allows you to include additional map services as background maps. Select BA_Detailed and LITGLStLouis and add them to the Selected Map Services as illustrated on the next page.

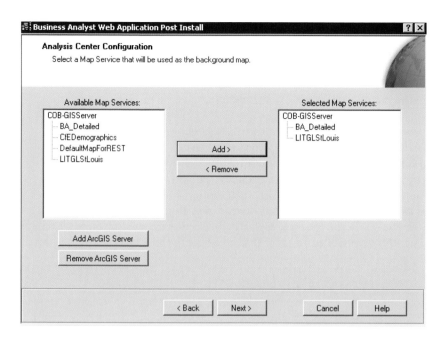

In the next window, which is illustrated below, you will select the ArcGIS Online Standard Services option and—from the services offered in the next window—you will select World Imagery to add a base layer of satellite imagery from ArcGIS Online to the project.

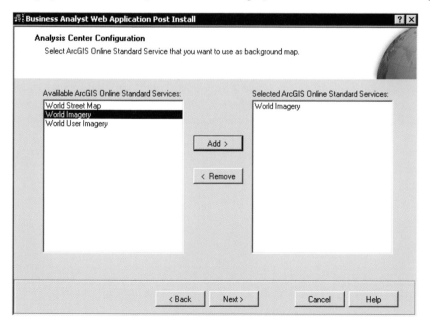

By clicking Next and accepting the default settings in the next two windows, you will add these map services to your Business Analyst Server application.

Upload the relevant Business Analyst Project to the Server

The next step in the process also begins in the Business Analyst Desktop installation on the server. Here you will add content to the project file by creating a store layer of LITGL's first store in the St. Louis area. You will also use the Segmentation Module to produce a Tapestry Segmentation Profile of the St. Louis CBSA. This will serve as the base profile to which customer profiles of individual stores will be compared.

Once these additions to the LITGL St Louis project are complete you will upload the project to Business Analyst Server. To do so, you will open the Business Analyst Project Explorer, right-click the LITGL St Louis project and click Upload project. The screen below illustrates this step.

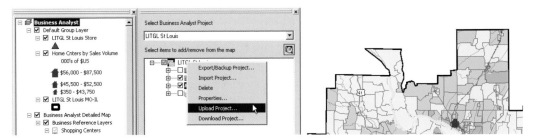

In the resulting Upload Project dialog box, you will enter the destination path in the Select Business Analyst Server box, then click Connect to open a connection to the server. You will then select a workspace in which to store the project and assign the project a name reflecting its content. When the Upload Project box resembles this you will click Upload to move the contents of the project file to the Business Analyst Server, where it will be available for users to access and use through a Business Analyst Server connection.

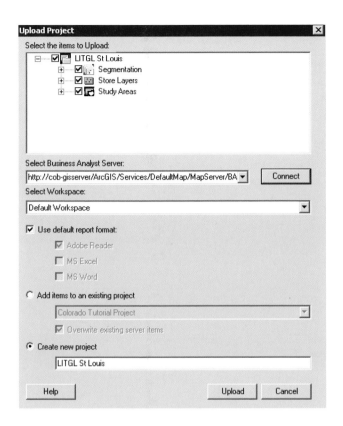

Manage security settings with ArcGIS Server and Business Analyst Server

Managing the security of Business Analyst Server resources serves several different objectives:

1. It protects enterprise data, as well as the company's investment in Business Analyst data, from exposure to all users on an open Web site.
2. It protects and allocates resource access internally, matching the resources available to individual users with their roles and responsibilities in the organization. This makes the system less complicated for users while allowing them to use the tools that support the decision-making tasks with which they are charged.
3. It allows system administrators to design a learning path for new managers in the organization. Initial tasks allow these managers to work with business GIS in a controlled, relevant project. This exposes them to the power of integrated business GIS tools and motivates them to develop skills relevant to more extensive use of these resources.

Your next task as system administrator is to create a new role, new store manager. This user role will be used by LITGL's newly hired store managers to generate Tapestry Segmentation Profiles of their customer base. This information will allow them to learn about the customers their store serves while, at the same time, developing the business GIS skills necessary to use the broader range of tools available to store managers.

Security management in Business Analyst Server is based on the settings, roles, and users established in the ArcGIS Server system. To manage security you will log on to the ArcGIS Server Manager system and select the Security tab. Your first tasks will be to enable security and create a database of ArcGIS Server roles and users. When you have completed that process, you will add roles and users to the system to manage access to ArcGIS Server resources. Business Analyst Server will build upon this system to manage access to Business Analyst resources more precisely.

You will add a new store manager role to the system by clicking Add Role from the Security page, entering New Store Manager as the Role name, and adding appropriate users to the role. The Add Role screen will resemble this.

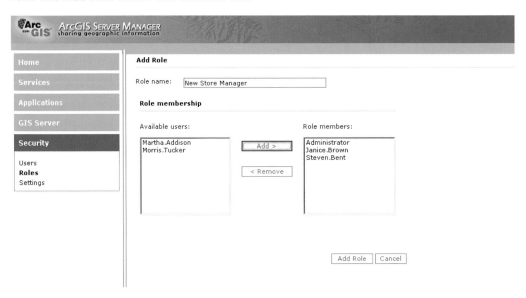

You must also create a User ID for Sarah Wilson, the new store manager in the St. Louis area. You will use the Add User procedure to create a user name for Ms. Wilson, specify her password as #Sx3m8L#, list her e-mail address, provide a password clue hint, and assign her to the new store manager role. The Add User screen will resemble the one on next page.

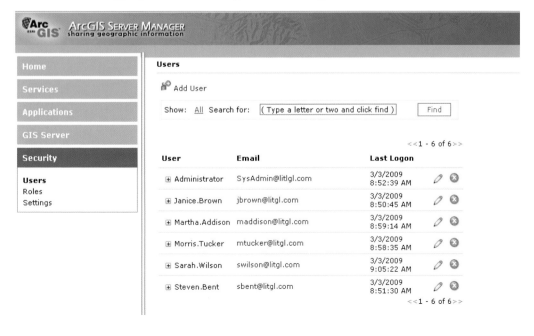

When this process is completed, the Users list will resemble the one below. Note that it contains your Administrator account, accounts for Janice and Steven as chief officers, and individual accounts for other users, including the new account for Sarah Wilson. By matching user accounts with system roles, you will assign each user the level of access to ArcGIS Server resources appropriate for their organizational responsibilities.

To manage additional security settings with Business Analyst Server you will run Business Analyst Web Applications Post Install and select the Configure Security option. You will then establish a connection to the database of roles and users you created in ArcGIS Server. The next step in the process will be to match each role with the specific tasks in Business Analyst Server that users assigned to that role will be able to access.

You will work with the Task Customization window illustrated below to assign the tasks necessary to create a Tapestry Segmentation Profile from a list of current customers to the new store manager role. Users in this role will be able to upload customer data, assign customers to stores based on an ID field, create a customer profile, and use it to generate selected reports, graphs, and maps. These 14 tasks include only a fraction of the Business Analyst Server tasks available, but are sufficient to produce very meaningful results for new store managers.

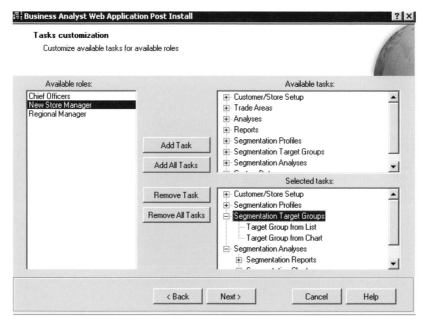

The other roles in the list in the left box are matched with different combinations of tasks. Essentially the chief officer's role can use all tasks, while the regional manager role is matched with tasks related to site selection projects, and the Store Manager role with tasks related to customer analytics. When Sarah Wilson improves her understanding of Business Analyst and her GIS skills, she will be granted the more extensive access to system resources available in the store manager role.

Standardize analytical procedures with workflows

With its Workflow concept, Business Analyst Server provides an additional resource for structuring users' analytical tasks. Workflows are step-by-step guides to standardized analytical procedures which guide users through tasks in the proper order.

Business Analyst Server provides two built-in workflows: Customer Analytics and Site Selection Evaluation. Customer Analytics is appropriate for the store manager role and Site Selection Evaluation Workflow is appropriate for the regional manager role.

Business Analyst Server also allows system administrators to design customized workflows and assign them to user roles. You will use this capability to create a Tapestry Segmentation Profile Workflow for the new store manager role.

To accomplish this task, you will log on to Business Analyst Server in your administrator role and open the Workflow Manager. To create a workflow, you will first design a Work-flow Template with a standardized set of procedures, then use the template to create a new workflow.

The screen below illustrates the process of designing templates, which are ordered collections of steps, each composed of discrete actions, maps, or reports. In this screen, you are creating a profile of customers from a data table. Review the actions already included in the template, through which you will add customer data from an external file and generate a Tapestry Seg-mentation Profile Report. Using this process, you will continue to add steps and actions to the template to produce reports and maps reporting the Tapestry Segmentation characteristics, buying and media behaviors of the customers of Sarah Wilson's store.

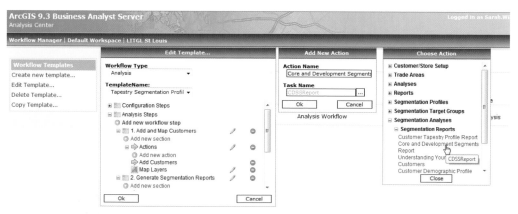

When you have completed the Workflow Template, you will use it to create a Tapestry Segmentation Profile Workflow. You will do so by clicking the Create Workflow option within Workflow Manager, naming the workflow and designating the template it will use as illustrated in the screen on page 280. The workflow is now complete and ready for use in the LITGL St Louis project in Business Analyst Server.

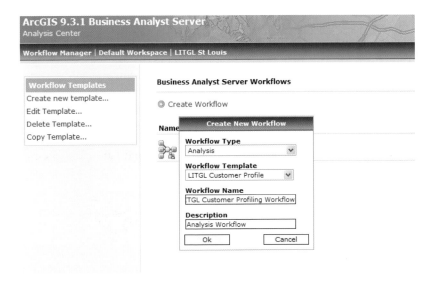

By performing these tasks as system administrator you have created a system user name for Sarah Wilson, defined her role relative to system resources, specified the Business Analyst Server tasks that she can use within that role, and created a workflow that will guide her through the task of generating demographic and Tapestry Segmentation Profiles of her customers.

Using BA Server applications with a Web browser

You will now switch perspectives and access the Business Analyst Server system as Sarah Wilson and explore the options available to her in her role as a new store manager.

Log on to the Analysis Center and review its structure, map contents and tasks

Sarah Wilson, store managers, and regional managers will access Business Analyst Server through the browsers on their desktop or laptop systems. As you simulate her role in this system, you will access the server hosting the BA Server application and log on using the user ID and password assigned in the security system above. Your first screen will be the Analysis Center, where you will select LITGL St Louis in the Project field. Your screen resembles the image below.

In this role, you have two options for initiating your Business Analyst Server session. The first, Business Analyst Tools, allows you to work directly with the Business Analyst capabilities enabled for you in the new store manager role. The second, LITGL Customer Profiling Workflow, guides you step-by-step through the process of uploading customer records to the system and producing a series of reports on their Tapestry Segmentation composition and behavior. To view the difference, click the Business Analyst Tools to open the initial Business Analyst Server screen, which resembles the one below. The initial map is in the window on the right. On the left are a group of vertical tabs each with an arrow at the right of the tab header with arrow symbols that allow users to expand or contract that panel. Use these symbols to expand and contract each tab to review their contents.

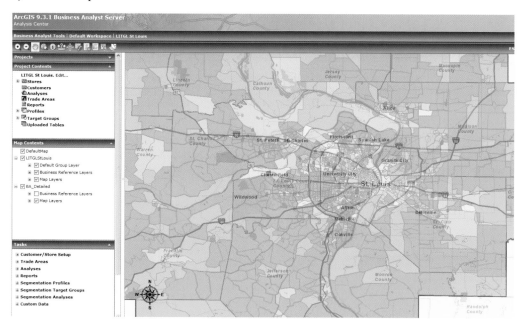

To familiarize yourself with the St. Louis market area, you wish to view the thematic maps of relevant customer characteristics. In Business Analyst Server, you do so by selecting the appropriate panel and adjusting layer visibility. In this way, you can adjust the Map Contents panel settings to display the location of LITGL's St. Louis store as well as the thematic maps you wish to see. Specifically, the Default Group Layer includes displays of LITGL's St. Louis store and the St. Louis CBSA. Business Reference Layers contains a layer of competing home centers and two thematic layers: Owner Occupied Housing Units and Median HH Income by Block Group. These layers are based on a shapefile similar to the one you created in chapter 3. Making these adjustments allows you to produce thematic maps like the one on the next page.

These layers provide you and other new stores' managers valuable information about the location of their stores and competing home centers as well as the demographic characteristics of the population in the store's market area.

Expanding each of the items in the Tasks panel produces a list of the tasks available to you in the new store manager role. Note that this role has 14 tasks that have been assigned by the system administrator in the procedure described earlier in this chapter.

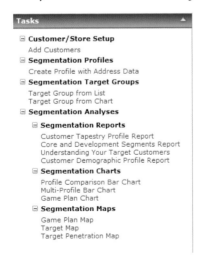

The small subset of tasks available to users in the Business Analyst Server system is appropriate for the new store manager role, which includes managers who are either new to the company and/or assuming managerial responsibilities for the first time. This limited range of tasks allows them to learn about their new stores and begin to use business GIS capabilities for the first time. As their knowledge and skills develop they will be reassigned to the store manager role, for which a much broader selection of tasks is available.

Use the LITGL Customer Profiling Workflow to upload customer data and create a segmentation profile and reports

To a new Business Analyst Server user, even the limited set of tasks available to new store managers can be daunting. While the task categories are well organized, the precise flow of tasks in an analysis is not clear to the novice user.

In this scenario, or in more advanced analyses by seasoned users, the Workflow model allows system administrators to integrate Business Analyst Server tasks in their proper order to produce desired analysis. In the role as system administrator above, you created a basic LITGL Customer Profiling Workflow for new store managers. You will now view that workflow in the role of Sarah Wilson.

Clicking Default Workspace in the menu bar at the top of the screen returns you to the Analysis Center. Click LITGL Customer Profiling Workflow to return to the default map and open the LITGL Customer Profiling Workflow. It contains a panel of Analysis Steps with three numbered tabs, each of which contains a series of tasks that you will perform in the order created by the workflow. The tasks in the tab labeled 1 allow you to create a customer layer from an MS Excel worksheet containing customer address and purchasing information. In the window below, you identify the source file, indicate the proper fields to be used in geocoding, and supply a name for the customer layer file.

11

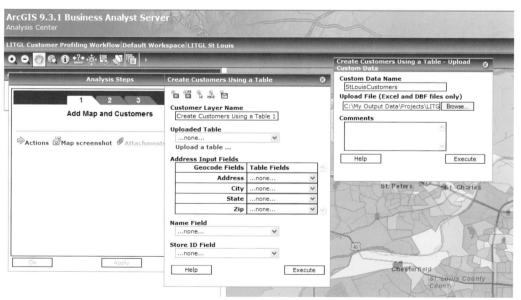

The addresses in the customer data table are geocoded and the resulting customer layer displays in the map. You can now observe the geographic distribution of the store's customers. Additional steps in the workflow allow you to learn more about them. Specifically, the steps in Tab 2 guide you through the process of creating a Tapestry Segmentation Profile from a customer table. You do so by clicking the action entitled Create Profile from Address Data to open its dialog box, naming the profile **St Louis Customer Profile**, confirming that St Louis Customers is selected as the Uploaded Table, enabling the Use Volume Information Field, and selecting LYPurchase as the Volume Information Field. These settings are illustrated in the window below. Clicking the Execute button runs the procedure with these settings.

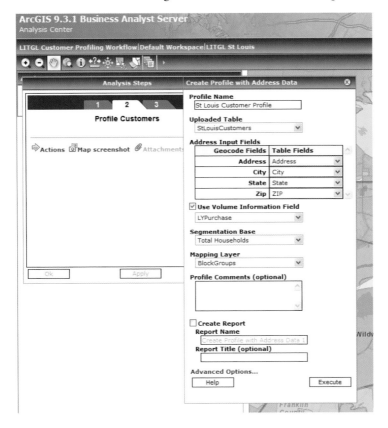

Business Analyst Server applies these settings, runs the procedure, and creates the customer profile layer you specified. It is now ready for use in generating reports on your customer base. As you continue the workflow, you will use this profile to generate demographic and Tapestry segmentation profile reports.

Clicking the Analysis Step 3 tab opens the Generate Reports window, which displays a list of the available reports. Clicking the Customer Tapestry Profile Report opens the dialog box for this task. The settings illustrated below create a report named **St Louis Customer Tapestry Report**, which uses St Louis CBSA as the Base Profile, St Louis Customer Profile as the

Target Profile, and displays the report as a PDF document. Clicking the Execute button runs the procedure with these settings.

Business Analyst Server generates the report, adds it in the Reports area of the workflow step, and displays it for you to review. When you scroll through the report to view the data on individual segments, the report will resemble this. It is identical in format to the report you created for the Minneapolis-St. Paul market area using the Segmentation Module in chapter 10. You will use a similar procedure to generate a Demographic Profile Report.

11

Customer Tapestry Profile by Segment

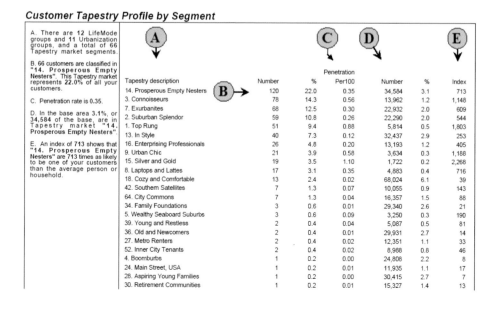

In the Analysis Step 3 tab, you will click Customer Demographic Profile Report to open its dialog box. Adjusting the settings to **LITGL St Louis Customer Demographic Profile** as the Report Name, BlockGroups as the Select geography level and LITGL St Louis Customers as the customer layer produces the window illustrated below. Selecting the View Report option and clicking the Execute button runs the procedure with these settings.

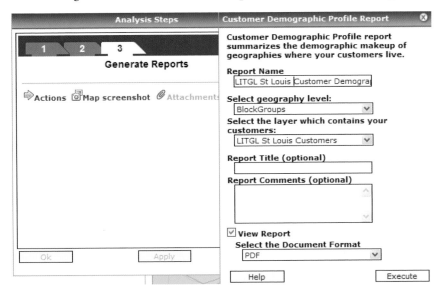

Business Analyst Server generates the following report and displays it on the screen.

Customer Demographic Profile

Who are your current customers?

Now that you know where most of your current customers are, you can learn more about their lifestyles and demographic makeup. ESRI uses the geographic information obtained based on your customer's address information, to append demographic characteristics and a Tapestry segmentation code to each customer record. The results of this analysis are found in the *Customer Demographic Profile and Customer Tapestry Profile*.

The Demographic Profile summarizes the demographic makeup of geographies where your customers live. Based on the information in this report, you can get a sense of the demographic makeup of your customer base.

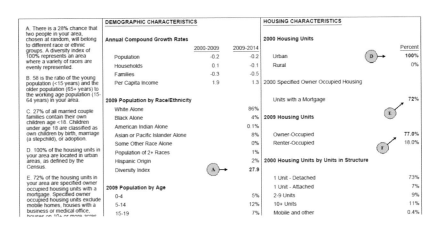

These profile reports reveal the essential demographic and Tapestry characteristics of LITGL's St. Louis customers. You may use additional steps in the workflow or explore the additional tools available to you in the Business Analyst Tools interface to expand your analysis of the St. Louis customer base.

These exercises explore a very modest selection of Business Analyst Server capabilities, though they do provide some insight into the power of the tools available in Business Analyst Server. Indeed, most of the applications you have used in this book are available in the Server environment as well. The default Customer Profiling and Site Selection Workflows offer comprehensive approaches to these projects which organize those applications in effective ways.[3] The ability to integrate enterprise data extends substantially beyond the table uploads performed here. And, significantly, these powerful resources are now accessible across the enterprise through browser technology, not dedicated GIS workstations. That said, your modest initial Business Analyst Server experience in the role of Sarah Wilson has been very productive. Through the Business Analyst Server mapping services you have learned about your competitive environment and the demographic characteristics of your market area. You have executed a standardized Workflow to geocode customer records, create a Tapestry Segmentation Profile Report and a Demographic Profile Report. The other tasks available in the New Store Manager role allow you to customize the Tapestry segments you wish to identify as Primary and Secondary targets, and create several reports to help you reach them.

The positive results of this initial session clearly illustrate the benefits of integrated business GIS analysis for your store and the value of the Business Analyst Server system in making them available. As you develop your skills with these tools more fully, you will be able to use the expanded capabilities available in the Store Manager role.

From a system administrator perspective, this chapter illustrates two levels of Business Analyst Server security management, which allow you to create learning paths for new managers seeking to develop business GIS skills. Standardized workflows, coupled with designated sets of tasks, allow you to guide new users and allow them to experiment with different tasks. Expanding these tools in roles for managers with more responsibility allows you to accommodate more accomplished users while matching roles and corporate duties as well. The full access available at the chief officer's level allows you and other senior managers to manage and assess that process as it develops.

Though not developed in detail here, you have also used the capabilities of ArcGIS Server to manage security, publish maps as Web services, and develop applications around them. Using these tools, you are able to use the ArcGIS Server framework to integrate these resources into enterprise information systems and Web applications for customers. Similarly, you can deploy map services for ArcGIS Explorer users or for use with corporate Web sites using the ArcGIS API for Javascript. Thus, these tools provide not only advanced business GIS processing across the enterprise, but a robust Web presence for LITGL in a variety of applications. With this integration, Living in the Green Lane has completed the process of integrating spatial analysis throughout its history and across its operations.

ROI considerations

There are two streams of ROI considerations relevant to the implementation of a Business Analyst Server installation. The first is the ability to apply integrated business GIS tools to decision making across the enterprise. From an ROI perspective, this is a question of the scalability of ROI calculations for each of the applications you have developed in this book. In other words, with each new expansion, profiling, merchandising, territory design, or routing application, the enterprise replicates the ROI impact discussed in that chapter and leverages its integrated business GIS investment even further. These benefits would be available to Living in the Green Lane no matter how it chooses to implement integrated business GIS on an enterprise-wide basis.

The second stream of ROI considerations focuses on the method of implementation. The basic choice here is between multiple single licenses of Business Analyst's suite of products installed at strategic points throughout the enterprise or an enterprise-wide Business Analyst Server installation.

In LITGL's case, the first option would involve Business Analyst Desktop and Segmentation Module installations at the regional or CBSA level, with corresponding costs of analysts for each system. In this scenario, successful analytical approaches could be replicated across the enterprise by sharing datasets, project files, and analytical techniques. This system would concentrate business GIS knowledge and skills within a group of desktop users working with the Business Analyst installations. Operating managers would depend on these analysts to perform requested analyses, report the results, and interpret their implications for decision making.

The alternative is an installation of Business Analyst Server for users across the enterprise. In this approach, Business Analyst functionality would be available directly to managers on both desktop and mobile devices using common browser interfaces. Managers would have access to datasets and procedures directly related to their operations and their geographic and managerial areas of responsibility. The integrity of data sets, standardization of common procedures, and management of system security would be managed centrally.

From an ROI perspective, the Business Analyst Server approach has significant tangible and intangible benefits over the multiple Business Analyst Desktop approach. The tangible benefits include substantial savings in hardware, software, and personnel costs. A Business Analyst Server installation includes ArcGIS, ArcGIS Server, the Business Analyst extension to ArcGIS, the Segmentation Module extension to Business Analyst, and the Business Analyst Server extension to Business Analyst and ArcGIS Server. This is an impressive suite of products. However, the acquisition cost of Business Analyst Server is only about 20 percent higher than the combined costs of a single Business Analyst Desktop and Segmentation Module installation. Thus, from a software perspective, Business Analyst Server costs more than one installation of its software components, but significantly less than two.

In the multiple installation option, the third installation and each subsequent one would magnify the cost advantage of Business Analyst Server.

These savings extend as well to the computer systems that house the software. While the servers for Business Analyst Server would be more expensive than each Business Analyst Desktop machine, each server would replace several desktop systems, creating a cost advantage. Further, the centralized system administration of the Business Analyst Server installation would produce savings of business GIS personnel costs as well.

In short, the tangible cost benefits of Business Analyst Server over multiple discrete installations including software, hardware, and personnel savings, begins with the second discrete Business Analyst Desktop installation, and grows substantially with each discrete installation beyond that.

The intangible benefits of the Business Analyst Server approach result from the enhanced business GIS skills of the managerial team, more knowledgeable and creative applications of business GIS tools by those managers, and the resulting greater exploitation of the decision support potential of business GIS across the enterprise. In this scenario, the enterprise-wide integration of business GIS skills geographically, functionally, and managerially creates a leaner, more flexible, and more responsive organization. Thus, these capabilities become a competitive advantage for Living in the Green Lane in its very dynamic, competitive business environment.

In short, Business Analyst Server offers significant tangible and intangible ROI benefits to organizations that manage it wisely. It also supports the enterprise-wide commitment to business GIS as a fundamental management tool that enables full exploitation of these tools. Finally, it strategically positions the enterprise to implement new integrated business GIS tools and applications as they emerge. The potential for such development is the focus of this book's conclusion in chapter 12. Before moving to that topic, however, take a moment to reflect on what you have learned in this chapter.

Summary of learning
You have expanded your GIS knowledge by learning:

1. The value of serving GIS resources across the enterprise
2. The value of developing managerial business GIS skills with distributed systems
3. The necessity of protecting enterprise and GIS data in distributed systems
4. The value of creating customized maps and datasets for business GIS applications and disseminating them across the enterprise
5. The value of establishing user access to appropriate resources through security systems defining users and roles
6. The value of enabling managers to access business GIS resources relevant to their responsibilities, apply them to enterprise data from their operations, and produce maps, reports, and charts that support effective decision making.

You have enhanced your GIS skills by using Business Analyst Server to:

1. Design maps for Web-based use and create map services from them
2. Use the ArcGIS Server security system to define user names and roles
3. Create and upload Business Analyst projects to Business Analyst services
4. Manage users and their access to business GIS resources with the Business Analyst Server security system
5. Manage access to existing workflows and create customized workflows
6. Access Business Analyst Server resources in a client role and upload data to the system
7. Perform analytical tasks and execute a customized workflow in a Business Analyst Server environment

Notes

1. Business Analyst Server can be installed in a desktop environment for testing and development purposes, but this configuration does not support full deployment. As are result, the graphic interface is determined by server side software and differs from the desktop interface, and so the images in this chapter resemble those you would see in a full scale installation rather than a desktop testing environment.

2. Visit the ArcGIS API for Javascript Resource Center at `http://resources.esri.com/arcgisserver/apis/javascript/arcgis/` to learn more about this technology.

3. To learn more about these workflows, click Default Workspace in the Business Analyst Server interface to reach the Start page. From here you can click the Overview of Business Analyst Workflows link to view a PDF document describing these two workflows in detail. In the Project box, you can select the Colorado Tutorial Project to review the Customer Analysis Workflow or the San Francisco Tutorial Project to review the Site Evaluation Workflow.

Chapter 12

Growth trajectories with integrated business GIS

This book has followed the journey of Living in the Green Lane from the original vision of its founders to its emergence as a national retailer. At every stage in that journey, the company has relied on the integrated business GIS technologies of ESRI Business Analyst for vital strategic information and analysis. The work you have performed has been an important foundation for the company's success.

With your help, Janice and Steven have successfully positioned LITGL for a bright future. The company has a proven business model whose attractiveness is poised to increase as health, wellness, and sustainability become more significant goals to consumers, businesses, and governments alike. Even in challenging economic times, each of these groups will increasingly seek out ways to function more efficiently and sustainably. Around the world, success in these efforts requires translating new environmental technologies into practical products for homeowners and businesses—a core competency of LITGL. As they prepare to exploit these opportunities, Living in the Green Lane's management team must determine the role business GIS will play in that effort.

That process should focus on two strategies:

- To identify opportunities to apply the current tools of Business Analyst more extensively within the company. New applications identified with this strategy have a direct beneficial impact on the ROI of ESRI Business Analyst for two reasons: First, they leverage the investment in Business Analyst resources that has already been made. Second, they enable more intensive and sophisticated use of both internal and Business Analyst data resources, producing even greater understanding of LITGL's customers and competitive environment. Thus, additional sales and decreased costs are obtained with minimal additional investment in business GIS resources.
- To seek opportunities for expanding business GIS resources beyond the scope of the current Business Analyst. These might include implementing new applications as they are added to the suite, investing in specialized GIS applications that extend or complement BA tools, and exploiting opportunities to integrate Business Analyst data and tools with other enterprise information technology resources.

Taken together, these strategies provide a comprehensive approach for identifying significant business GIS growth trajectories for Living in the Green Lane or any other organization.

Strategy one: leveraging the resources of the current ESRI Business Analyst suite of products.

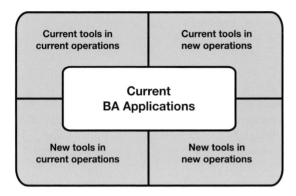

Figure 12.1 Growth trajectories within ESRI Business Analyst

Current tools in current operations

In this category of applications, Living in the Green Lane should continue to use the Business Analyst tools that have contributed so significantly to its success. In each of its markets, the search for new store locations can use the same trade area analysis, customer profiling, and site-selection techniques you have performed to date, maximizing the probability of success for each new location. These tools also will support the proper distribution of company-owned stores and franchised locations.

Vigorous expansion to new geographic markets will continue to rely on the powerful tools of the ESRI Segmentation Module to identify the best expansion opportunities across the country as well as the best locations within each potential new market region.

LITGL's growth also will be fueled by innovative new products and rapidly evolving consumer preferences and spending patterns. Business Analyst's customer profiling technologies, especially the tools of the Segmentation Module, Tapestry Segmentation system, and annual Market Potential index data will enable the company to track consumer trends and respond to them quickly.

New tools in current operations

This category of applications involves using new tools in Business Analyst to improve the company's performance in existing operations. In this context, the term "new tools" refers to existing components of Business Analyst that the company has not yet deployed in its planning processes. In many cases, these are more sophisticated implementations of BA features that the company has used. However, greater experience and an increasing body of customer and sales data bring the more advanced models into the company's range.

12

For example, you used basic Huff models on two occasions to estimate the market areas of competitors and to predict sales penetration of new stores. In these situations, you provided estimates of the factors that affect store attractiveness to potential consumers. As the body of sales data grows and you conduct survey research to determine customer preferences more precisely, more advanced versions of the Huff model become practical.

Specifically, you may use sales or survey data with the Advanced Huff Model with Calibration to create more sophisticated models of consumer preference for retail alternatives. These models will increase the accuracy of sales projections and patronage probabilities, which, in turn, will support better site selection decisions for future stores. As the number of LITGL stores grows and their markets become more saturated, this increased insight into the competitive environment becomes even more crucial.

Another application in this category is the development of consumer profiles based on Market Potential Index (MPI) survey data. In the Segmentation Module, you created profiles of current customers and several geographic areas for comparative purposes. These profiles, in turn, provided the foundational data for the segmentation analysis necessary to develop penetration and expansion strategies.

In contrast, survey profiles are based on MPI data. For example, to identify potential customers for Living in the Green Lane's organic lawn and garden maintenance services, you might create a profile of consumers who indicated in the MPI survey data that they "purchased organic soil additives in past 12 months." By comparing this profile with LITGL's customer base and the relevant service area, you would be able to assess the similarity of these prospects with current customers, the size of the group in the relevant marketing area, and its distribution relative to sales and service resources. This approach would provide invaluable information in product and service line decisions within Living in the Green Lane's marketing strategy. Tracking this data over time with annual survey updates would also help the company understand evolving consumer preferences.

Similarly, the accumulation of sales and customer data over time would allow you to exploit the potential of Territory Design more fully. Your original sales territory scheme was based largely on expenditure data and geographic considerations. As you accumulate data, you would shift the focus in this tool from potential expenditures to actual sales. This will allow you to balance sales potential more precisely to create more efficient and effective territories.

Current tools in new operations

This category of applications involves using current Business Analyst tools to support new marketing initiatives within the enterprise. For Living in the Green Lane such an initiative might involve creating ways to serve potential customers located outside the service areas of its retail and franchise stores. One such group would be potential customers who live in geographic areas where LITGL has stores but outside the service regions of those stores. Collectively these potential customers represent significant purchase volume, even though their visits to the company's stores are likely quite infrequent. They are also attractive prospects because they would be exposed to Living in the Green Lane's local advertising campaigns.

Thus they might be attracted to the company and its products, but disinclined to travel large distances to visit the stores.

To reach these potential customers, Living in the Green Lane might consider a targeted direct-marketing campaign with the objective of making high-value, low-frequency sales to them. You would use the Segmentation Module to identify concentrations of potential customers matching a desired profile. Highly focused newspaper inserts or mailings could offer incentives on items such as large appliances, and solicit subscriptions to e-mail or mobile phone notification services for promotional pricing on items of interest. This group might also be served with online ordering and pickup services, which would bring these customers to stores to pick up items ordered online. While there, of course, they would be motivated to browse the centers for other attractive items. This approach, based on the power of the Segmentation Module to pinpoint attractive customers, would increase LITGL's customer base and sales in the geographic areas where it has a retail presence.

New tools in new operations

This category of applications involves using new Business Analyst tools to support new marketing initiatives. For Living in the Green Lane this might mean expanding the Web-based GIS tools you worked with in ESRI Business Analyst Server to the company's broader e-commerce site. This approach would be similar to the direct-marketing strategy just discussed but with a much broader audience. That initiative would target customers in geographic areas where the company has a retail presence. The e-commerce initiative would target customers anywhere but with greatest attention to those areas in which Living in the Green Lane has not established stores.

Your work in Business Analyst Server was largely focused on disseminating BA data and capabilities to managers throughout the enterprise. The e-commerce initiative envisioned here clearly incorporates Web-based systems beyond that scope. However, BA Server and other ESRI technologies such as ArcWeb Services; ArcGIS Online; ArcGIS Server; and the ArcGIS API for JavaScript, Flash, and Silverlight, support the spatial dimension of the broad e-commerce implementation.

And that spatial dimension should be considerable. Traditionally, it would include such functions as finding the nearest store, providing directions and/or tracking orders. In LITGL's case, however, several other applications would be of considerable value. In evaluating payback periods for investments in wind, solar, and insulation technologies, consumers require climate information. By integrating this data from existing map services and customizing it for the consumer's geocoded location, LITGL's spatially enabled site would help consumers make the right product choices for their area. A buffering operation using BA tools and drive-time data would identify the nearest contractors capable of installing desired products. Scheduling and tracking tools would allow customers to schedule installation conveniently and monitor the status of deliveries to meet that schedule.

12

Taken together, these technologies, many of them based on BA data and tools, would allow Living in the Green Lane to extend its reach quite effectively to customers outside the reach of its retail centers.

Strategy two: implementing business GIS beyond the current Business Analyst suite

Business GIS solutions extend beyond the current Business Analyst suite of products and offer significant potential for improving Living in the Green Lane's strategic planning, managerial decision making, and operational efficiency. The potential approaches in this strategy include implementing new business GIS applications as they become available in Business Analyst, deploying specialized business GIS solutions that extend the power of Business Analyst and integrating Business Analyst with Enterprise GIS resources to extend the power of these technologies across the enterprise.

Figure 12.2 depicts these approaches, which are discussed in more detail below.

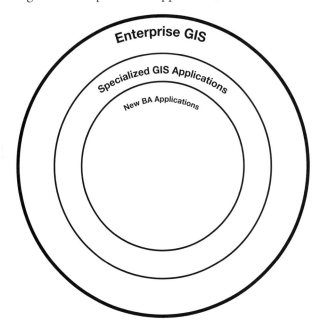

Figure 12.2 Growth trajectories beyond current Business Analyst applications

New Business Analyst applications

As Business Analyst develops, each new release adds additional functionality and data to the system. Often these innovations are available first in the Business Analyst Online system, especially as new datasets and report formats become available. However, major releases of Business Analyst typically include new automated models, analyses, and reports, as well.

For example, geocoding and routing procedures are continually being improved, while both the Segmentation Module and Territory Design extension have been incorporated into the system. Often these applications are a general implementation of more detailed solutions crafted by ESRI and its business partners. As these applications are added, they extend the functionality of Business Analyst and increase the system's ROI potential.

The most valuable applications of this type include analyses that are inherently spatial, of general value to a variety of enterprises, and subject to standardized solutions with automated wizard-based tools. Two such potential new applications are Business Continuity Planning and Sustainability Assessment and Reporting.

Business Continuity Planning refers to a broad set of tools whose overall objective is ensuring that enterprises can operationally survive natural, accidental, or man-made disasters. This is a form of contingency planning for both security concerns and natural disasters in an increasingly risky business environment. These applications can be as simple as mapping alternative exit routes for employees and customers in an organization's physical facilities. More complex solutions involve preparing for industrial accidents or natural disasters, whose effects are generally localized and whose impacts are relatively minor. Even more complex are responses to planned, coordinated attacks on facilities, which might target communication and transportation infrastructure assets as well as physical facilities.

In each of these situations, Business Continuity Planning encompasses emergency response procedures as well as rapid implementation of contingency plans to preserve enterprise function and operations. The inherently spatial nature of these problems and the necessity of integrating environmental, infrastructure, climate, and emergency response data into the planning process make integrated business GIS systems a valuable tool for Business Continuity Planning.

Sustainability Assessment and Reporting is a response to the increasing pressure on organizations of all kinds to assess and manage their environmental impact in a comprehensive framework. The pressure, in turn, emerges from heightened awareness that environmental impact goes well beyond the point source pollution of manufacturing facilities and that it has long term consequences for climate and environmental quality.

In this environment, managing for sustainability has both economic and social implications. Economically, it provides opportunities for companies to streamline operations, reduce energy consumption, and deploy resource-saving strategies for more efficient operations. Socially, it provides opportunities for Living in the Green Lane to align itself with important social goals while performing its economic functions more efficiently.

Sustainability Assessment and Reporting is an attractive integrated business GIS application for several reasons. First, it is also an inherently spatial enterprise. The location of facilities, employees, suppliers, distribution systems, and customers all impact the resource consumption of the enterprise. Moreover, assessing the impact of serving these constituencies requires the type of flexible analysis available in the Business Analyst framework.

12

Second, this process requires models that merge data from several different sources. Climate and location impact the feasibility of renewable energy resources and green-design techniques. Transportation infrastructure and the availability of public transport affect potential programs to reduce travel and/or delivery costs. The availability and accessibility of alternative supply resources affects the organization's ability to improve its product design and production efficiency. Integrated business GIS systems can access these data sources and incorporate them into its analytical models.

Third, although some concerns apply across organizations, this process also must be customized to the unique situation of each enterprise based on its industry, location, operational methods or enterprise culture. By integrating the data sources just discussed with internal enterprise data, integrated business GIS systems empower organizations to approach this process in a manner consistent with its own philosophy and objectives.

As Business Analyst grows to incorporate these and other applications, it provides enterprises such as Living in the Green Lane the ability to address significant new issues within the context of a familiar system. This, in turn, enables companies to address these issues more quickly and at less cost while leveraging even further their investment in integrated business GIS technology.

Specialized GIS applications

The business GIS network involves not only software companies such as ESRI, but a wide range of business partners, many of whom build specialized solutions on an ArcGIS platform. Organizations with substantial operations in the areas targeted by these applications can benefit greatly from their increased power relative to the basic tools provided in Business Analyst. Examples of this pattern include routing applications and logistics management.

Business Analyst provides basic routing capabilities for sales and service calls. You used this resource to solve a routing problem in chapter 8. However, the specialized routing and dispatch systems described in that chapter allow their effectiveness to be extended significantly. Vehicles equipped with GPS allow for real-time monitoring of vehicle position and progress by dispatchers. Integration of real-time weather, traffic, road repair, and accident-reporting map services allow dispatchers to integrate this data, reroute personnel, and very rapidly distribute new routing information to vehicles and mobile devices.

For enterprises with substantial sales, delivery, and/or service operations, these specialized systems multiply the ROI of routing systems dramatically and pay for themselves very quickly. Within the ESRI product line, ArcLogistics provides many of these capabilities, while other business partners provide more specialized solutions customized to the needs of each enterprise.

While these systems focus on outbound delivery and service schedules, other partners provide comprehensive logistics-management packages covering both inbound and outbound transportation as well as facilities scheduling and inventory management solutions. These systems provide a comprehensive approach for managing the flow of goods into and out of manufacturing, distribution, and retail facilities. They provide similar types of efficiency and cost benefits as do routing systems but on a larger, more comprehensive scale.

Integration with enterprise GIS

GIS resources are rapidly becoming significant components of enterprise information systems through a variety of mechanisms. The most common format is desktop GIS installations with access to enterprise data. Increasingly, however, other information systems access and use GIS data resources in their own configurations.

One example is the customer relationship management software used by many companies to capture comprehensive data on customers, their purchase histories, and buying preferences. Integrating spatial data into these systems allows companies to include this information in scheduling and routing systems where, for example, it might affect the priority of sales or service calls.

More comprehensively, Enterprise Resource Planning (ERP), Business Intelligence (BI), and IS developers such as Oracle incorporate some tools for spatial analysis within their systems. This can be the result of proprietary GIS tools in their applications, integration of these tools in the development of enterprise systems using resources such as ArcObjects, and the use of resources such as ArcGIS Online, ArcGIS Server, or ESRI Business Analyst Server services to add GIS tools to enterprise systems. This development is facilitated through the ArcGIS APIs for JavaScript, Flash, Silverlight, and through Business Analyst Online, which allow developers to use Web-based map services and geoprocessing tools in conjunction with enterprise resources to author Web-based mashups of applications customized to the needs of enterprises and their constituencies.

However these systems are implemented, they create the opportunity to access enterprise information resources with Business Analyst tools on the one hand and share business GIS tools across the enterprise through Business Analyst Server technology on the other.

A comprehensive map for business GIS growth

Figure 12.3 provides a comprehensive map of these two strategies and the potential growth trajectories they create for business GIS within Living in the Green Lane's enterprise development model.

12

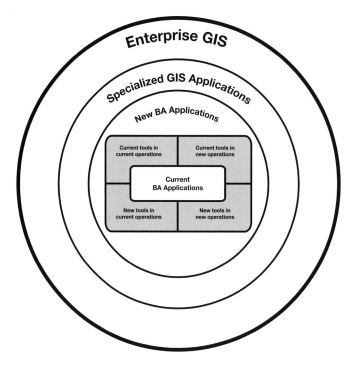

Figure 12.3 Integrated business GIS expansion map

Although this model emerges directly from Living in the Green Lane's situation, it has general value for enterprises implementing integrated business GIS systems with Business Analyst. The specific sequence of deploying BA applications will vary from organization to organization. Even so, the growth trajectories of fully exploiting BA's potential while also seeking out new, specialized, and enterprise-level business GIS solutions are sound opportunities for most enterprises.

Equally significant, this model also provides a growth strategy for your acquisition of business GIS knowledge and skills. In this book, you have increased your understanding of integrated business GIS and its value in decision making. You have also developed your skills in the Business Analyst applications that help enable a new enterprise to emerge from a founding vision and grow into a successful national organization. The growth trajectories in figure 12.3 offer you a learning program as you seek to extend your understanding and skills in this vital technology.

It has been a pleasure to guide you on the earliest stages of your journey. Good luck on the rest of your exciting odyssey.

Select market potential indexes for Living in the Green Lane's high purchases segment

Market Potential Indexes: Attitude, Media, and Activity Behaviors	Tapestry Segmentation Segment					
	02	04	06	07	12	13
Apparel						
Bought hiking/backpacking apparel last 12 months	103	144	126	143	132	80
Bought running apparel in last 12 months	152	176	139	142	113	92
Bought athletic shoes in last 12 months	119	129	117	113	108	121
Spent $75+ on athletic shoes in last 12 months	140	156	131	105	133	126
Attitudes						
Consider self very conservative	106	104	96	134	111	113
Consider self somewhat conservative	142	142	118	143	127	129
Consider self middle of the road	97	96	115	98	92	103
Consider self somewhat liberal	139	101	125	102	103	118
Consider self very liberal	75	86	82	97	95	82
Automobile						
Rented truck/trailer in last 12 months	106	125	130	135	144	117
Rented truck to move personal/household goods last 12 months	113	124	129	128	145	117
Household has navigational system in vehicle	154	188	120	123	122	115
Civic activities						
Participated in any public activity in last 12 months	124	118	119	123	105	113
Made contributions to PBS in last 12 months	176	126	121	166	71	141
Participated in environmental group in last 12 months	149	142	115	161	87	107
Recycled products in last 12 months	145	134	137	134	109	125
Member of civic club	177	101	83	202	70	105
Lawn and garden						
Purchased lawn-maintenance service in last 12 months	195	209	129	165	146	135
Used service for property/garden maintenance last 12 months	219	215	138	164	127	143
Spent $150+ on property/garden maintenance in last 12 months	194	176	152	176	137	137

Table A   Select market potential indexes for Living in the Green Lane's high purchases segment

Purchased garden insecticide in last 12 months	163	143	132	163	122	125
Purchased lawn insecticide in last 12 months	176	187	146	168	144	145
Purchased organic soil additives in last 12 months	142	163	132	204	90	136
Grocery						
Drank bottled water/seltzer in last 6 months	118	120	112	109	103	108
Used bran bread in last 6 months	123	88	94	89	90	99
Used multigrain bread in last 6 months	154	145	134	165	118	131
Used oat bread in last 6 months	129	121	120	124	83	118
Used wheat bread in last 6 months	113	123	113	109	117	109
Used white bread in last 6 months	88	95	98	93	98	92
Used breakfast/granola/fruit bars & snacks in last 6 months	132	133	128	115	128	124
Used fish/seafood (fresh or frozen) in last 6 months	113	106	107	112	97	104
Used fresh fruit/vegetables in last 6 months	104	102	104	105	102	104
Health						
Exercise at home 2+ times per week	131	132	121	133	108	126
Exercise at club 2+ times per week	175	180	136	132	128	134
Exercise at other facility (not club) 2+ times per week	107	131	124	128	113	144
Buy foods specifically labeled as natural/organic	145	134	115	92	101	114
Used vitamin/dietary supplement in last 6 months	121	114	116	118	101	117
Used last 12 months: SPF 15+ suntan/ sunscreen product	151	156	149	142	129	132
Home improvement and services						
Household used professional carpet cleaning service in last 12 months	178	193	149	166	140	138
Houshold used professional exterminator in last 12 months	152	162	105	141	136	128
Used housekeeper/maid in last 12 months	166	162	105	130	106	119
Used housekeeper/maid/professional household cleaning service in last 12 months	173	164	122	138	117	130

Table A   Select market potential indexes for Living in the Green Lane's high purchases segment *(cont.)*

Internet						
Use Internet more than once a day	170	181	139	138	133	142
Internet last 30 days: made personal purchase	191	190	155	147	131	145
Internet last 30 days: obtained real estate information	160	209	134	122	157	153
Internet last 30 days: made travel plans	208	209	154	150	135	156
Ordered anything on Internet in last 12 months	177	172	151	149	139	145
Leisure activities/lifestyle						
Did bird watching in last 12 months	117	94	118	175	102	135
Cooked for fun in last 12 months	126	139	117	139	117	115
Did furniture refinishing in last 12 months	117	129	137	164	84	104
Did painting/drawing in last 12 months	118	111	113	88	89	96
Did photography in last 12 months	147	130	156	149	106	134
Read book in last 12 months	131	132	123	142	112	126
Surfed the Internet in last 12 months	147	156	135	131	128	135
Played video game in last 12 months	82	123	104	95	115	117
Did woodworking in last 12 months	123	85	122	132	109	111
Mail and phone orders/Yellow Pages						
Ordered any item by mail/phone in last 12 months	140	121	123	139	115	122
Spent on mail orders in last 12 months: $200+	124	123	121	103	90	112
Spent on phone orders in last 12 months: $500+	157	154	161	188	110	152
Media						
Heavy radio listener	84	87	99	94	100	88
Radio format listen to: classic hits	115	129	152	128	157	130
Radio format listen to: classic rock	120	118	134	100	116	109
Radio format listen to: classical	214	98	108	139	63	149
Radio format listen to: jazz	139	129	125	124	104	107
Radio format listen to: news/talk	177	156	159	168	107	170
Radio format listen to: public	185	165	97	158	68	138
Heavy magazine reader	127	117	115	108	112	112
Read epicurean magazines	197	152	107	127	93	121

Table A   Select market potential indexes for Living in the Green Lane's high purchases segment *(cont.)*

Read gardening magazines	85	71	129	98	92	82
Read health magazines	117	112	101	110	103	100
Read home-service magazines	136	130	122	116	112	110
Read music magazines	59	82	81	60	94	79
Read science/technology magazines	159	101	126	106	93	94
Read travel magazines	175	166	124	132	110	117
Heavy newspaper reader	138	115	108	132	101	110
Read newspaper: food/cooking section	132	105	116	138	100	122
Read newspaper: home/furnishings/ gardening section	145	123	133	141	112	122
Read newspaper: sports section	119	119	111	118	106	107
Read newspaper: travel section	159	141	129	149	109	134
Sports						
Participated in aerobics	141	169	131	150	134	132
Participated in bicycling (mountain)	184	195	124	127	135	136
Participated in bicycling (road)	139	154	162	136	110	133
Participated in fishing (fresh water)	89	83	99	106	95	97
Participated in jogging/running	153	190	144	142	142	131
Participated in martial arts	50	128	89	92	90	93
Participated in racquetball	125	186	96	73	104	133
Participated in roller blading/in-line skating	90	156	142	127	118	141
Participated in swimming	149	146	136	138	124	118
Participated in snorkeling/skin diving	188	144	160	150	104	163
Participated in tennis	174	198	133	107	126	127
Participated in walking for exercise	142	135	124	142	119	131
Participated in weight lifting	147	171	142	140	132	121
Participated in yoga	149	128	121	124	135	85
Travel						
Domestic travel—vacation/honeymoon in last 12 months	140	134	125	127	118	125
Went backpacking/hiking domestic vacation in last 12 months	166	171	163	180	179	98
Went to beach on domestic vacation in last 12 months	173	151	154	145	114	152
Visited National Park on domestic vacation in last 12 months	182	130	153	158	139	139

Table A   Select market potential indexes for Living in the Green Lane's high purchases segment *(cont.)*

Spent on domestic travel in last 12 months: $3,000+	239	210	145	205	95	158
Played golf on domestic vacation in last 12 months	239	216	154	211	96	235
Foreign travel—vacation/honeymoon in last 3 years	178	162	137	144	103	136
Member of any frequent flyer program	224	213	155	175	147	151
Source: Mediamark Research, ESRI Tapestry Segmentation Demonstration CD, 2007						

Table A   Select market potential indexes for Living in the Green Lane's high purchases segment *(cont.)*

Installing the LITGL Minneapolis St. Paul CBSA project folder; customer support resources

Importing the project data folder

Once you have ArcGIS Desktop 9.3.1, Business Analyst Desktop 9.3.1 and the Segmentation Module on your system, You may proceed to the following steps for importing the LITGL Minneapolis St Paul CBSA project folder into Business Analyst Desktop:
(This step should be done only after installing all necessary components, licensing the software, and activating the data):

1. Using Windows Explorer, copy the project file, LITGL Minneapolis St Paul.zip, onto your hard drive from the book's enclosed DVD or its Web page at **www.esri.com/esripress**. Your desktop might be a good place to place the zip file, but do not unzip this file.
2. Open Business Analyst from Start > All Programs > ArcGIS > Business Analyst > Business Analyst.
3. Open the Project Explorer window by clicking on the button shown below:

4. Under Select items to add/remove, right click on Default Project and choose Import Project.
5. Navigate to the folder where you saved the zipped file in step 1 above. Select the LITGL Minneapolis St Paul.zip and choose Open.
6. It will take several seconds to extract the files from the archive.
7. Under Select Business Analyst Project, click on the drop down arrow and choose LITGL Minneapolis St Paul.

You are now ready to begin the tutorials!

Checking for Business Analyst updates

If you have access to the Internet, you can check for updates to Business Analyst by:

1. Selecting this item in the Business Analyst program menu.
2. From the Start menu, go to ArcGIS > Business Analyst > Check for Business Analyst Updates.
3. When the dialog box appears, click the Check button to see if there are any available updates.
4. If updates are available, you can choose to download them and also run the updates after they are downloaded.
5. For more information, click the Help button on the dialog box.
6. Please note that to check for Business Analyst updates, ArcMap and ArcCatalog must be closed. If you are unable to run the automatic update utility in Business Analyst due to local security protocols, you can download the latest service patch from the support site at **http://support.esri.com** and run the patch locally.

Additional information and resources

ArcGIS Business Analyst Help can be found in the ArcGIS Desktop Help topics listed on the Contents tab under Extensions. For help using the geocoder in Business Analyst, more information can be found in your ArcGIS Business Analyst\Documentation folder.

ESRI, 380 New York Street, Redlands, CA 92373-8100
909-793-2853
Fax 909-793-5953
`www.esri.com`

Basic Instructions for loading Business Analyst 9.3.1

C:\Program Files\ArcGIS\Business Analyst\Documentation\install_guides

Tips for resolving startup issues

C:\Program Files\ArcGIS\Business Analyst\Documentation\Troubleshooting_Guide.pdf

Creating an incident

There are two ways you can submit requests for support:
1. Call directly for support at 888-377-4575, when prompted, press 2.
2. Create an online request for support at `http://support.esri.com/`. This method will enable you to fill out a detailed form, as well as attach any supporting documents.

Training resources

Using ArcGIS Business Analyst Manual.
`http://resources.arcgis.com/content/product-documentation?fa=viewDoc&PID=78 &MetaID=1083`

Additional help for using Business Analyst:
`http://webhelp.esri.com/arcgisdesktop/9.3/index.cfm?id=3774&pid=3767& topicname=Getting _ additional _ help _ for _ Business _ Analyst`

Additional help for learning the Segmentation Module:
`http://resources.arcgis.com/content/product-documentation?fa=viewDoc&PID=78 &MetaID=1304`

Additional help for learning the Territory Design Module:
`http://resources.arcgis.com/content/product-documentation?fa=viewDoc&PID=78 &MetaID=1304`

User forums

`http://forums.arcgis.com/forums/46-Business-Analyst`

Business Analyst blog
`http://blogs.esri.com/Dev/blogs/businessanalyst/default.aspx`

Fred L. Miller, PhD, is Thomas Hutchens Distinguished Professor of Marketing and Business GIS in the Department of Management, Marketing, and Business Administration at Murray State University, Murray, Kentucky. He is also director of MSU's Regensburg Exchange Programs. His teaching and research interests are in the fields of business GIS, e-commerce, emerging technologies in marketing, and global marketing management. Miller authored the book *GIS Tutorial for Marketing* (ESRI Press, 2007).

Data source credits

Screenshot Data Sources

Data in Business Analyst screenshots is courtesy of ESRI; Tele Atlas North America, Inc.; InfoUSA, Inc.; Directory of Major Malls, Inc.

Custom Data

\ChapterFiles\Chapter2\SA.shp, created by the author
\ChapterFiles\Chapter3\LITGLBusinessPlan.mxd, created by the author
\ChapterFiles\Chapter4\AvailableProperties.xls, created by the author
\ChapterFiles\Chapter4\LITGLFirstStore.mxd, created by the author
\ChapterFiles\Chapter5\LITGLFirstStore.mxd, created by the author
\ChapterFiles\Chapter6\LITGLCustomers.mxd, created by the author
\ChapterFiles\Chapter6\LITGLCustomers.xls, created by the author
\ChapterFiles\Chapter7\LITGLExpansion.mxd, created by the author
\ChapterFiles\Chapter8\LITGLSalesMgt.mxd, created by the author
\ChapterFiles\Chapter8\LITGLSalesMgtRoute.mxd, created by the author
\ChapterFiles\Chapter9\LITGLCustomersFull.xls, created by the author
\ChapterFiles\Chapter9\LITGLCustomersFull.xls.xml, created by the author
\ChapterFiles\Chapter9\LITGLProfile.mxd, created by the author
\ChapterFiles\Chapter10\LITGLCustomersFull.xls, created by the author
\ChapterFiles\Chapter10\LITGLCustomersFull.xls.xml, created by the author
\ChapterFiles\Chapter10\LITGLSegmentation.mxd, created by the author
\ChapterFiles\Chapter10\LITGLSegmentation2.mxd, created by the author
\ChapterFiles\Chapter11\StLouisCustomers.xls, created by the author
\ChapterFiles\Chapter11\StLouisCustomers.xls.xml, created by the author
\MapFiles\Available Properties.lyr, created by the author
\MapFiles\BGIncEdOwnExp.shp, created by the author
\MapFiles\Ch10CustomersByStore.shp, created by the author
\MapFiles\Ch3MinnStPaulSA.shp, created by the author
\MapFiles\Ch4HomeCenters.shp, created by the author
\MapFiles\Ch4HomeCenters1.shp, created by the author
\MapFiles\Ch5AvailableProperties.shp, created by the author
\MapFiles\Ch5CustomerProspecting.shp, created by the author
\MapFiles\Ch5HuffEqlProb.shp, created by the author
\MapFiles\Ch5ThresholdRings.shp, created by the author

\MapFiles\Ch6DriveTimeTradeAreas.shp, created by the author
\MapFiles\Ch6LITGLStore.shp, created by the author
\MapFiles\Ch7AvailableProperties.shp, created by the author
\MapFiles\Ch7HomeCentersGreenData.shp, created by the author
\MapFiles\Ch7LITGLCustomers.shp, created by the author
\MapFiles\Ch7MasonSiteSA.shp, created by the author
\MapFiles\Ch7SteiersDriveTime.shp, created by the author
\MapFiles\Ch8LITGLStores.shp, created by the author
\MapFiles\Ch8SalesRepHomes.shp, created by the author
\MapFiles\Ch8SalesSeedPoints.shp, created by the author
\MapFiles\Ch8ServiceCalls.shp, created by the author
\MapFiles\Ch8ServiceCalls.xml, created by the author
\MapFiles\Educational Attainment by Block Group.lyr, created by the author
\MapFiles\Home Centers by Sales in 000's.lyr, created by the author
\MapFiles\Home Improvement Expenditure Grids.lyr, created by the author
\MapFiles\Home Ownership by Block Group.lyr, created by the author
\MapFiles\Home Related Expenditures per HH.lyr, created by the author
\MapFiles\Lawn and Garden Expenditures by ZIP.lyr, created by the author
\MapFiles\LITGL Customer Prospecting.lyr, created by the author
\MapFiles\LITGLFirstStore.lyr, created by the author
\MapFiles\LITGLMinnStPaul.lyr, created by the author
\MapFiles\LITGLStores.lyr, created by the author
\MapFiles\Median Home Value Block Group.lyr, created by the author
\MapFiles\Median Household Income by Block Group.lyr, created by the author
\MapFiles\Service Calls.lyr, created by the author
\MapFiles\ZIPIncEdOwnExp.shp, created by the author

Segmentation
\Target groups\Lifemodes\metadata.xml, standard Business Analyst target groups
\Target groups\Lifemodes\TargetGroup.xml, standard Business Analyst target groups
\Target groups\Urbanization\metadata.xml, standard Business Analyst target groups
\Target groups\Urbanization\TargetGroup.xml, standard Business Analyst target groups

Territories
\Territories\popbalterritories\tdlayer.gdb\FrozenTerritories, created by the author
\Territories\popbalterritories\tdlayer.gdb\level_2, created by the author
\Territories\popbalterritories\tdlayer.gdb\level_3, created by the author
\Territories\popbalterritories\tdlayer.gdb\Overlapped, created by the author
\Territories\popbalterritories\tdlayer.gdb\StorageTable, created by the author
\Territories\popbalterritories\tdlayer_bak.gdb\FrozenTerritories, created by the author
\Territories\popbalterritories\tdlayer_bak.gdb\level_2, created by the author
\Territories\popbalterritories\tdlayer_bak.gdb\level_3, created by the author
\Territories\popbalterritories\tdlayer_bak.gdb\Overlapped, created by the author
\Territories\popbalterritories\tdlayer_bak.gdb\StorageTable, created by the author
\Territories\popbalterritories\tdlayer.lyr, created by the author

Index

Related titles from ESRI Press

GIS Tutorial for Marketing

ISBN: 978-1-58948-079-7

GIS Tutorial for Marketing demonstrates the types of analysis that can broaden the impact and improve the efficiency of marketing efforts. Filled with scenario-based, hands-on exercises, this tutorial addresses topics like targeted promotional campaign development, merchandise strategy planning, retail site selection, and more. No previous GIS experience is required—this book will give beginning students or professionals the knowledge and experience required to gain a distinctive edge in planning marketing strategies and solving marketing problems.

GIS Tutorial 2: Spatial Analysis Workbook, Second Edition

ISBN: 978-1-58948-258-6

Updated for ArcGIS 10, *GIS Tutorial 2* offers hands-on exercises to help GIS users at the intermediate level continue to build problem-solving and analysis skills. Inspired by the ESRI Guide to GIS Analysis book series, this book provides a system for GIS users to develop proficiency in various spatial analysis methods, including location analysis; change over time, location, and value comparisons; geographic distribution; pattern analysis; and cluster identification.

Modeling Our World: The ESRI Guide to Geodatabase Concepts, Second Edition

ISBN: 978-1-58948-278-4

Modeling Our World presents a complete survey of the geodatabase information model. This book explains how to use geodatabase structural elements to promote best practices for data modeling and powerful geographic analyses; how to use rules and data properties in the geodatabase to ensure spatial and attribute integrity; how to manage your organizations work flow; how to scale geodatabases from small projects up to multiple departments across a large organization.

The Business Benefits of GIS: An ROI Approach

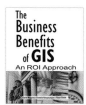

ISBN: 978-1-58948-200-5

The Business Benefits of GIS offers a fact-based, benefits-focused methodology aimed at ensuring the sustainability of GIS initiatives by effectively demonstrating the success of the investment. This book demonstrates a step-by-step framework with supplemental case studies, interactive digital tools, and templates that allow the reader to apply the book's methodology to GIS initiatives, and achieve consensus among stakeholders.

ESRI Press publishes books about the science, application, and technology of GIS.
Ask for these titles at your local bookstore or order by calling 1-800-447-9778. You can
also read book descriptions, read reviews, and shop online at www.esri.com/esripress.
Outside the United States, contact your local ESRI distributor.